Welding processes handbook

Welding processes handbook

Second edition

Klas Weman

Oxford Cambridge Philadelphia New Delhi

Published by Woodhead Publishing Limited,
80 High Street, Sawston, Cambridge CB22 3HJ, UK
www.woodheadpublishing.com

Woodhead Publishing, 1518 Walnut Street, Suite 1100, Philadelphia, PA 19102-3406, USA

Woodhead Publishing India Private Limited, G-2, Vardaan House, 7/28 Ansari Road,
Daryaganj, New Delhi – 110002, India
www.woodheadpublishingindia.com

First edition 2003, Woodhead Publishing Limited and CRC Press LLC
Second edition 2012, Woodhead Publishing Limited

British Library Cataloguing in Publication Data
A catalogue record for this book is available from the British Library.

Library of Congress Control Number: 2011934935

ISBN 978-0-85709-510-7 (print)
ISBN 978-0-85709-518-3 (online)

The publisher's policy is to use permanent paper from mills that operate a sustainable forestry policy, and which has been manufactured from pulp which is processed using acid-free and elemental chlorine-free practices. Furthermore, the publisher ensures that the text paper and cover board used have met acceptable environmental accreditation standards.

Printed by Lightning Source

Contents

Preface

Production of this guide to welding was originally prompted by a desire for an up-to-date reference on applications in the field. The content has been chosen so that it can be used as a textbook for European or International welding courses in accordance with guidelines from the European Federation for Welding, Joining and Cutting (EWF) and International Institute of Welding (IIW). Over the last years, an equivalent Swedish guide has been used for courses on welding processes and equipment. The author hopes that this guide will serve as a useful reference book for those involved in the welding field.

In writing the book, there has been a conscious effort to ensure that both text and illustrative material is clear, concentrating particularly on interesting and important aspects.

Although the book has been written in Sweden, with input from Swedish experts, it reflects technology and methods that are internationally accepted and used. My thanks are due to all those who have been involved in the work, with particular mention to:

Claes Olsson, HighTech Engineering, who has written the chapter on design of welded components.
Claes-Ove Pettersson, Sandvik, who has edited the section on stainless steel.
Curt Johansson, formerly SAQ, who has written the chapter on quality management.
Gunnar Lindén, Air Liquide, who has edited the chapter on welding costs.
Staffan Mattson, Aluminiumförlaget, who has written the chapter on welding of aluminium.

Curt Johansson sadly died during preparation of this new edition. The book is dedicated to him.

Klas Weman

1 Introduction to welding

1.1 The history of welding

Methods for joining metals have been known for thousands of years, but for most of this period the only form of welding was forge welding by a blacksmith.

A number of totally new welding principles emerged at the end of the 19th century. It became possible to combine and store safely gases such as oxygen and acetylene to produce a flame with enough heat. At the same time sufficient electrical current could then be generated for resistance welding and arc welding. The intensity of the heat source enabled heat to be generated in, or applied to, the workpiece quicker than it was conducted away into the surrounding metal. Consequently it was possible to generate a molten pool which solidified to form the unifying bond between the parts being joined. The basic welding methods of resistance welding, gas welding and arc welding were all developed during the time before World War I.

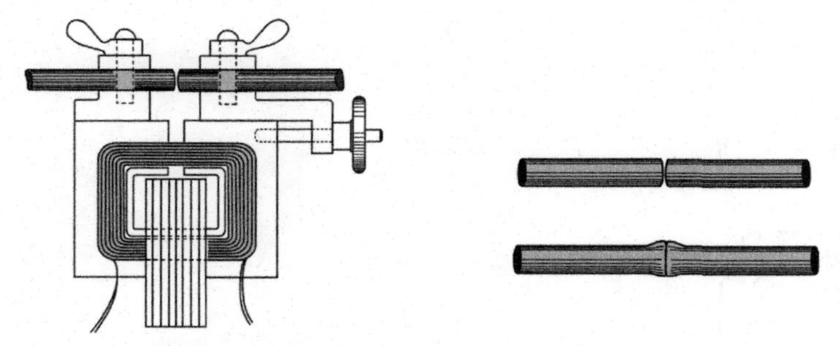

Figure 1.1 Thomson's resistance welding transformer.

Resistance welding

The first resistance welding machines were used for butt welding. Elihu Thomson made the first welding transformer in 1886 in the USA and patented the process the following year. His transformer managed to output approximately 2000 A at 2 V open circuit voltage.

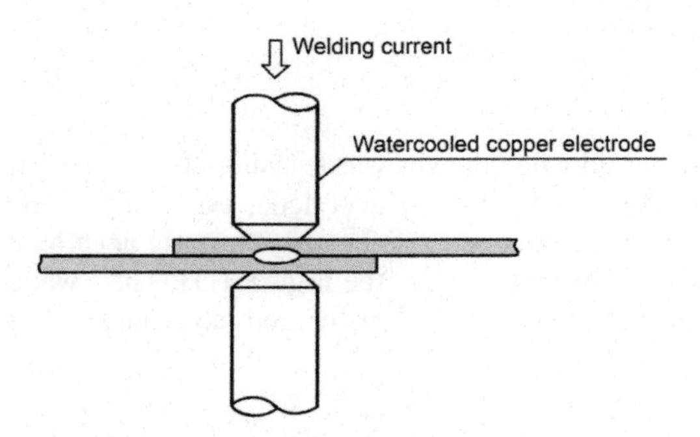

Figure 1.2 The principle of resistance spot welding.

After the turn of the century Thomson continued to develop other types of machines such as *spot welding, seam welding, projection welding* and *flash butt welding.* Spot welding later became the most common resistance welding method and today is used mainly in the automotive industry but also for many other sheet metal applications. Figure 1.2 shows the principle: two copper electrodes apply a pressure to the overlapping plates and a high current melts the plates together. The first robots for resistance spot welding were delivered from Unimation to General Motors in 1964.

Gas welding

Gas welding with an oxyacetylene flame was developed in France at the end of the 19th century. The first torch suitable for welding was made by Edmund Fouche and Charles Picard in about 1900. The use of acetylene and oxygen made it possible to produce a comparatively high flame temperature, 3100°C, which is higher than that of other hydrocarbon based gases. The torch became the most important tool for welding and cutting of steel.

Acetylene gas had been discovered much earlier in England when Edmund Davy in England found that a flammable gas was produced when carbide was decomposed in water. The gas proved to be excellent for illumination when burned and this soon became the main use of acetylene.

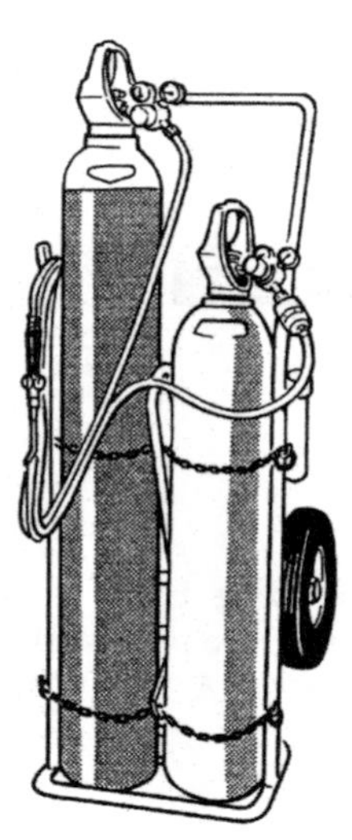

Figure 1.3 Equipment for gas welding.

However, numerous large explosions occurred when the gas was transported and used. It was found that acetone could dissolve large quantities of acetylene, especially if the pressure was increased. In 1896 Le Chatelier developed a safe way of storing acetylene by the use of acetone and a porous stone inside a cylinder (see Figure 1.3). The Swede Gustaf Dahlén at AGA changed the composition of the porous content and managed to make acetylene storage 100 % safe.

Arc welding

Arc welding was initially carried out using carbon electrodes, developed by Bernardos, and then by the use of steel rods. The weld was however not protected from the air and there were problems with quality. The Swede Oskar Kjellberg made an important advance when he developed and patented the coated electrode. The welding result was amazing and arc welding using coated electrodes became the foundation of the ESAB welding company.

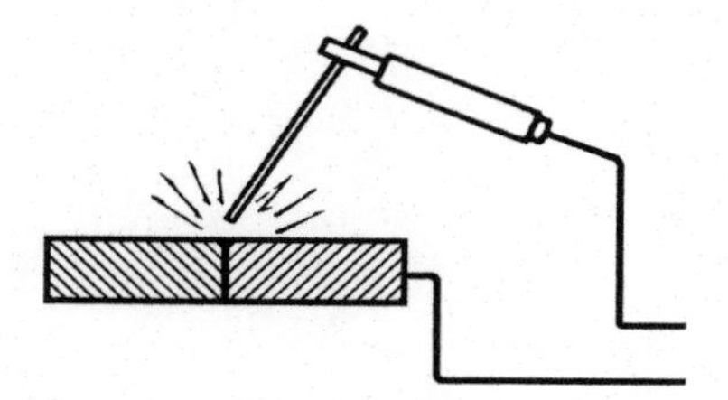

Figure 1.4 Principle of Manual Metal Arc (MMA) welding.

Later developments: SAW, TIG and MIG/MAG welding

Later, in the 1930s, new methods were developed. Up until then, all metal-arc welding had been carried out manually. Attempts were made to automate the process using a continuous wire. The most successful process was *submerged arc welding* (SAW) where the arc is "submerged" in a blanket of granular fusible flux.

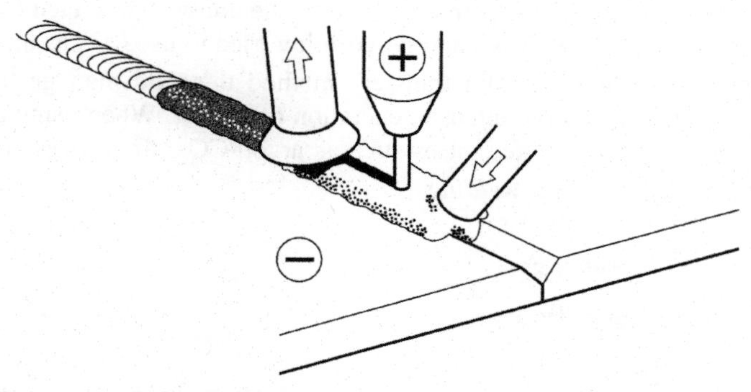

Figure 1.5 Submerged arc welding.

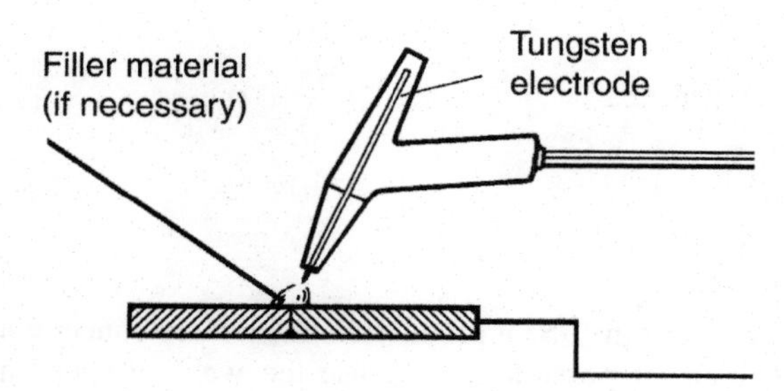

Figure 1.6 The TIG welding method.

During the Second World War the aircraft industry required a new method for the welding of magnesium and aluminium. In 1940 experiments began in the USA with the shielding of the arc by inert gases. By using an electrode of tungsten, the arc could be struck without melting the electrode, which made it possible to weld with or without filler material. The method is called Tungsten Inert Gas or *TIG welding*.

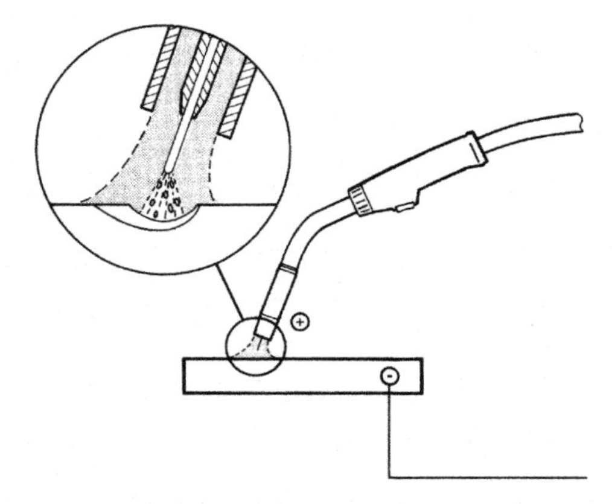

Figure 1.7 The MIG/MAG welding method.

Some years later the Metal Inert Gas or *MIG welding process* was also developed using a continuously fed metal wire as the electrode. Initially, the arc was protected by inert gases such as helium or argon. Lyubavskii and Novoshilov tried to use CO_2 as this was much easier to obtain. By using the "dip transfer" method they did manage to reduce some of the problems caused by the intense generation of spatter. When using a relatively reactive gas such as CO_2 or mixed gases such as argon/CO_2, the process is generally called Metal Active Gas or *MAG welding*.

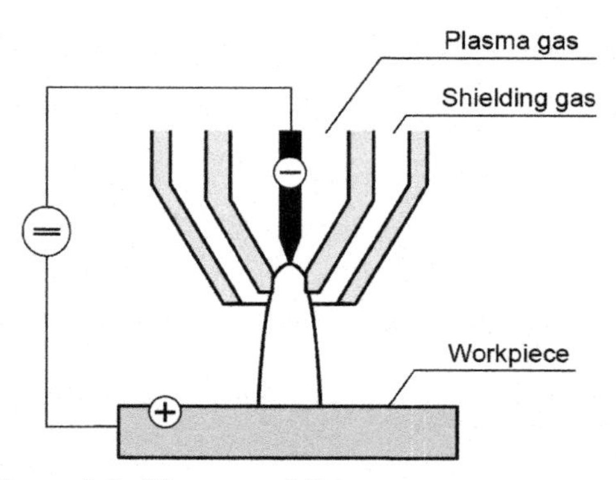

Figure 1.8 Plasma welding.

When *plasma welding* was introduced in 1953 it produced a much more concentrated and hotter energy source which made it possible to increase the welding speed and decrease the heat input.

Other welding processes

The power-beam processes, *electron beam* (EB) *welding* and *laser welding,* have the most intense heat sources. The breakthrough with EB welding came in 1958. The aircraft and nuclear power industries were the first to utilise the method. The main characteristics of EB-welding is its deep and narrow penetration. Its one limitation is the need for a vacuum chamber to contain the electron beam gun and the workpiece.

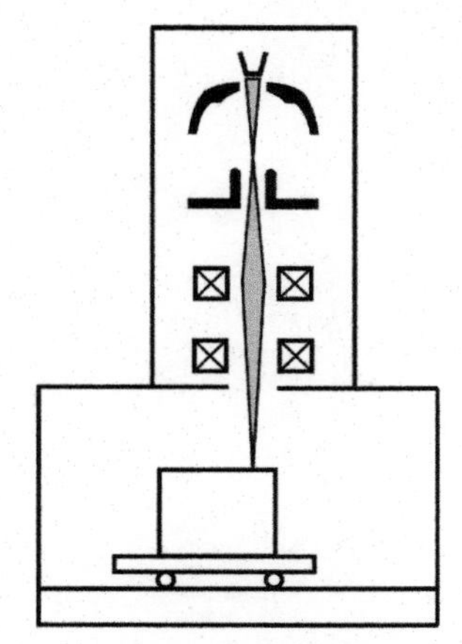

Figure 1.9 Electron beam welding.

In some respects, *laser welding* (and cutting) have ideal characteristics. The laser beam is a concentrated heat source which permits a high operating speed and very low distortion of the workpiece. Unfortunately, a high power laser is also large and expensive. The beam must also be conducted to the joint in some way. The light from a CO_2 laser must be transmitted by mirrors, while that from a Nd:YAG-laser can be carried by a thin glass fibre, which makes it attractive for use with robotic welding.

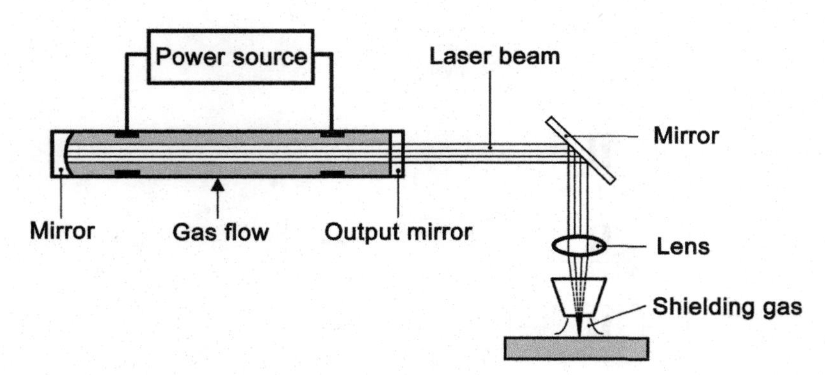

Figure 1.10 Laser welding.

Friction stir welding (see Figure 1.11) was patented in 1992 by TWI in the UK. The method works very well for aluminium and manages to join two pieces without melting and at high quality. It has no need of filler metal and has a low consumption of energy.

Another benefit is the low environmental impact. The process is so simple and so effective that it must be considered to be one of the 20th century's most remarkable innovations in the world of welding.

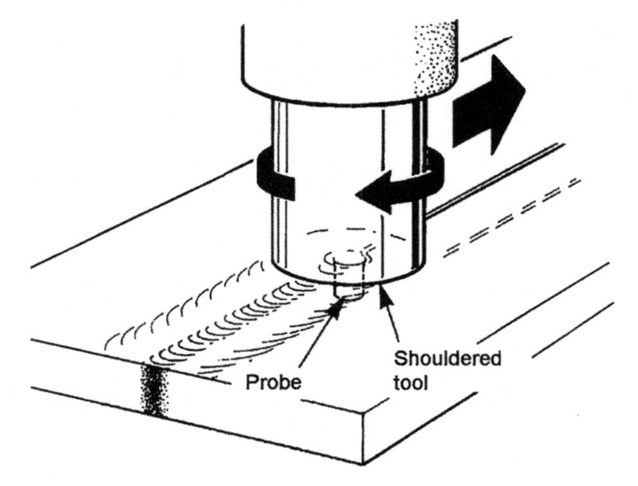

Figure 1.11 Friction stir welding (FSW).

The most recent development is *hybrid welding* where two different processes are combined. Particularly promising is laser-MIG hybrid welding where very high speed and high penetration are achieved.

Welding power sources

One possible reason why electric welding methods were not developed before the end of the 19th century may have been the lack of suitable power sources.

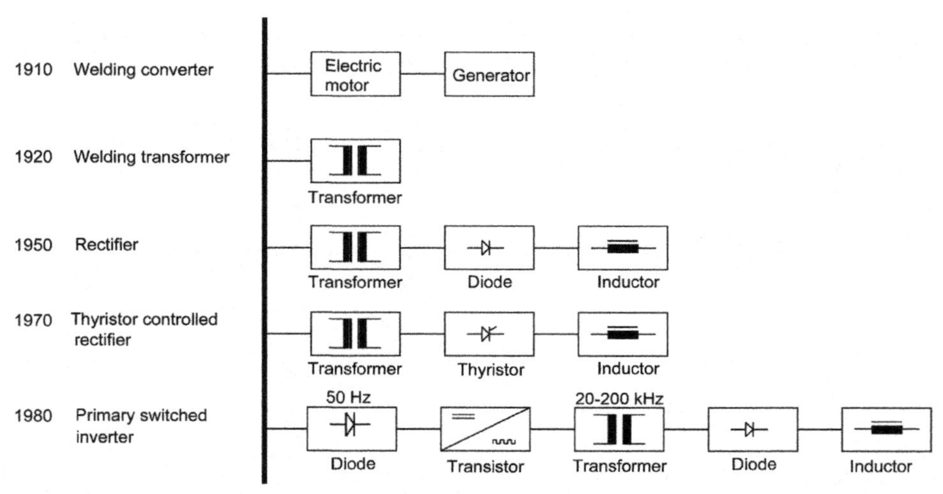

Figure 1.12 Different design principles of power sources.

AEG in Germany developed a *welding generator* in 1905. It was driven from a three phase asynchronous motor and had suitable characteristics for welding. The weight was 1000 kg and it developed 250 A.

DC current was normally used for arc welding until the 1920s. The development of stick electrodes made it possible to use AC current. The welding transformer soon became very popular, as it was cheaper and had lower energy consumption.

At the end of the 1950s static *welding rectifiers* were introduced. Selenium rectifiers were used initially, followed later by silicone rectifiers.

Thyristor rectifiers that came later made it possible to control the welding current electronically. This type of rectifier has since become very common at least for larger welding power sources.

However, the most remarkable development in welding power source started when the *welding inverter* was developed in 1980.

Mechanised welding

Robots have been used for resistance welding since 1964. The breakthrough with arc welding robots came about 10 years later. New electric robots could then be designed with enough high accuracy to satisfy the demands of MIG welding.

Mechanised welding opened up for new applications. Submerged arc welding became an important process for welding in ship yards, pipe mills and big pressure vessels. Narrow gap welding saved time, consumables and reduced distortion in welding of heavy sections. Mechanised TIG and plasma welding are both used in high quality applications.

1.2 Terminology

Definitions of welding processes

Definitions of welding processes are given in ISO 857. Reference numbers for the various processes are given in ISO 4063. Examples of the use of reference numbers for welding processes are shown in Table 1.1. These numbers are used as references in preparing drawings (ISO 2553) or in welding procedure specifications (EN ISO 15614-1).

TABLE 1.1 Reference numbers for some fusion welding processes (ISO 4063).

Welding method	Reference number
Metal-arc welding with coated electrode	111
Flux-cored wire metal-arc welding without gas shield	114
Submerged arc welding	12
MIG welding	131
MAG welding	135
MAG welding with flux-cored wire	136
TIG welding	14
Plasma arc welding	15
Oxy-fuel gas welding	311

A summary of the different types of welding processes is shown in Figure 1.13.

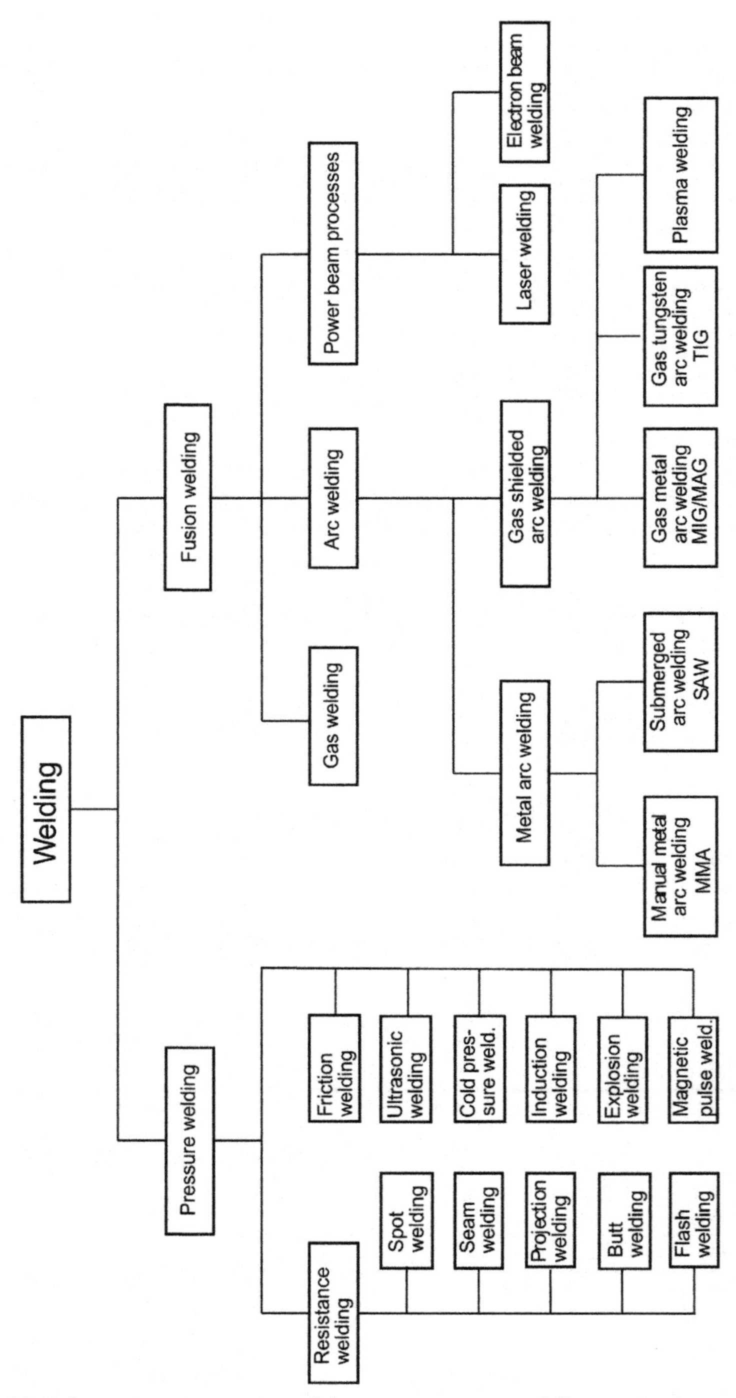

Figure 1.13 Schematic presentation of the most common welding methods.

Basic terms

Some basic terms used in welding are defined below.

Weldment. The combined weld, heat affected zone (HAZ – see below) and base metal.

Butt welds. Butt welds join two pieces fitted edge to edge. Full penetration welds are most common and provide a particularly strong weld.

Fillet welds. The alternative to the butt weld, fillet welds join two overlapping pieces (lap joint) or two pieces placed perpendicularly to each other (e.g. a T or L-shaped joint)

Pressure welding. Welding in which sufficient outer force is applied to cause more or less plastic deformation of both the facing surfaces, generally without the addition of filler metal. Usually, but not necessarily, the facing surfaces are heated in order to permit or to facilitate bonding.

Fusion welding. Welding without application of outer force in which the facing surface(s) must be melted. Usually, but not necessarily, molten filler metal is added.

Welding procedure specification (WPS). A document specifying the details of the required variables for a specific application in order to assure repeatability (EN ISO 15609).

Deposition rate. The amount of metal supplied to the joint per unit time during welding.

Parent metal. The metal to be joined, or surfaced, by welding, braze welding or brazing.

Longitudinal direction. The direction along the length of the weldment, parallel to the weld.

Transverse direction. The direction along the width of the weldment, perpendicular to the weld.

Surfacing. Producing a layer of different metal by welding, e.g. with higher corrosion, abrasion or heat resistance than the parent metal.

Heat input. The heat input has great importance for the rate of cooling of the weld. It can be calculated from the formula:

$$Q = \frac{U \cdot I \cdot 60}{V \cdot 1000} \cdot \text{Efficiency}$$

Efficiency*
MMA: 0.8
MIG/MAG: 0.8
SAW: 1.0
TIG: 0.6

where Q = heat input (kJ/mm)
U = voltage (V)
I = current (A)
V = welding speed (mm/min)

(* Efficiencies according to EN 1011-1.)

Heat Affected Zone (HAZ). The heat affected zone, (Figure 1.14), is that area of the base metal not melted during the welding operation but whose physical properties are altered by the heat induced from the weld joint.

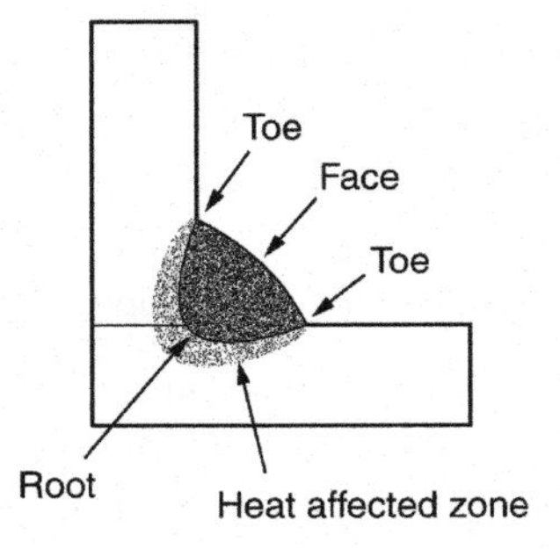

Figure 1.14 Fillet weld showing the location of weld toes, weld face, root and heat affected zone.

Throat thickness. Fillet welds are calculated by reference to the throat size. The size required is specified on drawings in terms of throat thickness, t, or the leg length, l, see Figure 1.15.

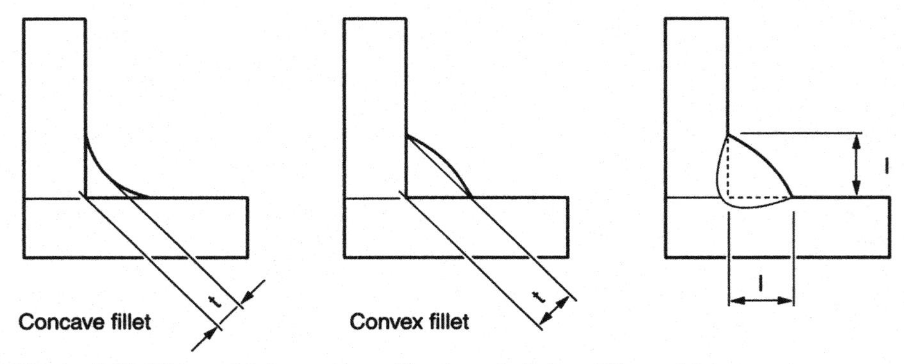

Figure 1.15 Throat thickness (t) and leg length (l) in a fillet weld.

Joint types and welding positions

Joint types are chosen with regard to the welding method and plate thickness. The ideal joint provides the required structural strength and quality without an unnecessarily large joint volume. The weld cost increases with the size of the joint. A higher heat input may cause problems with impact strength and distortion.

Joint preparation can be expensive. It is preferable to use joint types where the joint faces are parts of the workpiece. This means that fillet welds are probably the most commonly used joints.

Figure 1.16 illustrates some key terms in joint technology whilst Figure 1.17 shows examples of some common joint types.

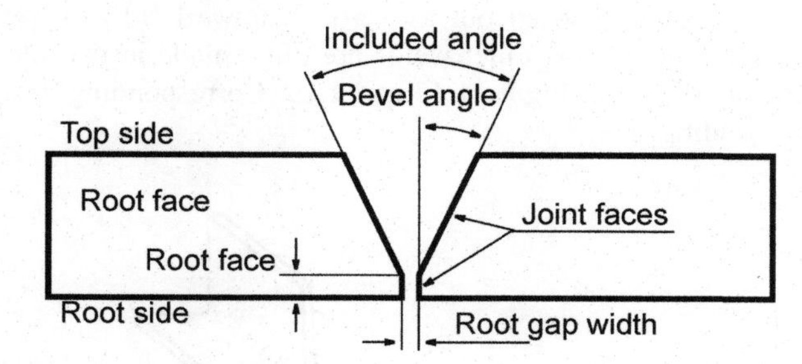

Figure 1.16 Joint terminology.

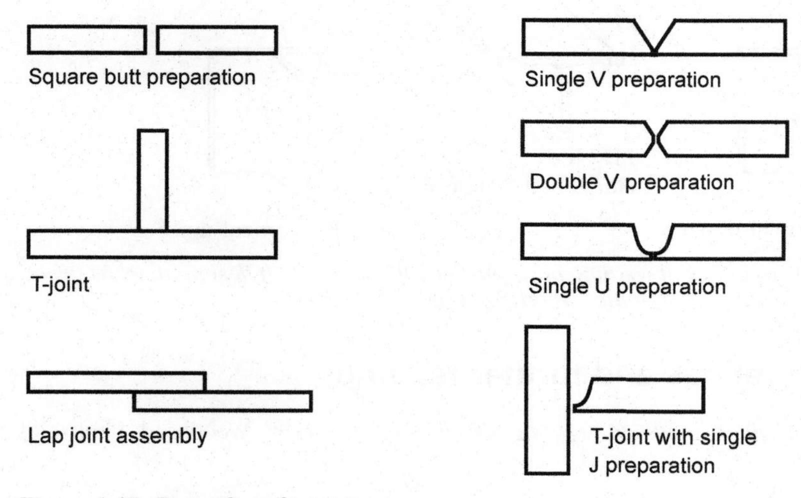

Figure 1.17 Examples of joint types.

Welding positions

There are essentially six different basic welding positions:

- PA, flat
- PB, horizontal vertical
- PC, horizontal
- PD, horizontal overhead
- PF, vertical up
- PG, vertical down

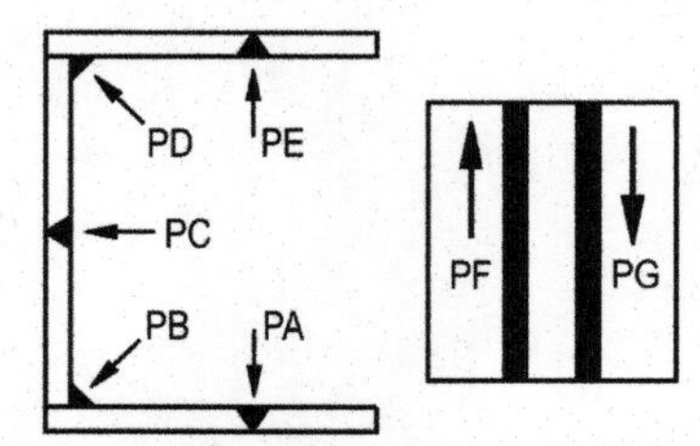

Figure 1.18 Designations for welding positions.

Vertical position welding can be carried out as vertical upward (PF) or vertical downward welding (PG), see Figure 1.18. Fillet welds are often made in the intermediate positions PB and PD, as shown in Figure 1.18 and 1.19. Corresponding designations are also used in pipe welding.

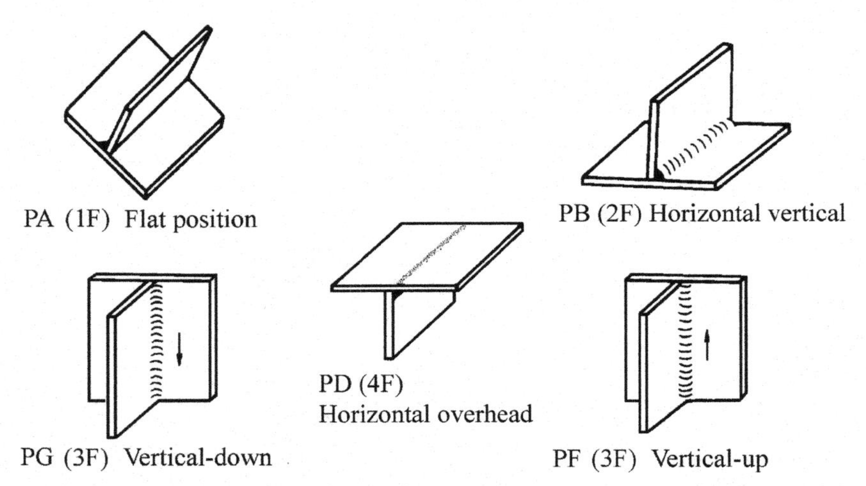

Figure 1.19 Definitions of welding positions for fillet welds with designations as given in EN ISO 6947. AWS designation in parenthesis.

1.3 References and further reading

S. Hughes, *A quick guide to welding and weld inspection*, Woodhead Publishing Limited, 2009.

K. Weman, *A history of welding*, Svetsaren, Vol. 59 No 1, 2004.

2 Gas welding

2.1 Introduction

Oxy-acetylene gas welding is one of the oldest methods of welding and, for many years, was the most widely used welding technique. Its use is a lot less common today. Nevertheless, it is a versatile method, using simple and relatively cheap equipment. It is suitable for repair and construction work, for welding pipes/tubes and structures with a wall thickness of 0.5–6 mm and in materials particularly prone to cracking, such as cast iron. It is also used for welding non-ferrous metals and for cladding and hardfacing. In addition to welding, the technique is often used for cutting, and is also very useful for heating and straightening materials.

The heat is generated by the combustion of acetylene in oxygen, which gives a flame temperature of about 3100 °C. This is lower than the temperature of an electric arc, and also produces a less concentrated heat. The flame is directed onto the surfaces of the joint, which melt, after which filler material can be added as necessary. The melt pool is protected from air by the reducing outer zones of the flame. The flame should therefore be removed slowly when the weld is completed.

The less concentrated flame results in slower cooling, which is an advantage when welding steels that have a tendency to harden, although it does make the method relatively slow, with higher heat input and the added risk of thermal stresses and distortion.

2.2 Equipment

A set of equipment for gas welding consists essentially of gas bottles, pressure regulators, gas hoses, flashback arresters and welding torches (Figure 2.1).

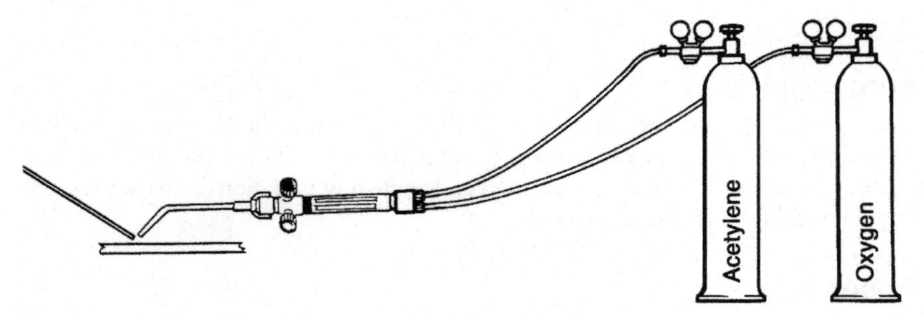

Figure 2.1 A gas welding set.

Welding gases and their storage

Gas bottles for combustible gases must be stored outdoors or in a well-ventilated area. Special warning signs must be displayed on the outside of the storage area. Acetylene and oxygen bottles must be kept well apart.

Acetylene

Acetylene (C_2H_2) is the main fuel gas for gas welding. Its main properties compared to other fuel gases are listed in Table 2.1. It consists of 92.3 % carbon and 7.7 % hydrogen by weight. Its combustion in oxygen produces a higher combustion temperature than that of any other hydrocarbon gas. In addition, its flame is the most concentrated in comparison with other gases.

It is important to be aware of the risks of using acetylene. Acetylene ignites very easily, and produces an explosive mixture in air over a wide range of concentrations (2.3–82 %). As a result it is important to check carefully that there are no leaks from gas bottles or hoses.

Acetylene is chemically unstable under pressure, even without the presence of air. Under certain conditions, it can explosively decompose to its constituents (carbon and hydrogen). To enable the gas to be stored safely, the bottles are filled with a porous mass, saturated with acetone, which absorbs the gas when it is filled. The pressure in the bottles is 2 MPa. However, explosive decomposition can occur in pipes or hoses leading from the bottle if the pressure in them exceeds 1.5 MPa.

TABLE 2.1 Important characteristics of fuel gases.

Gas	Density, kg/m^3	Calorific value, MJ/kg	Flame temperature, °C	Combustion velocity, m/s
Acetylene	1.07	48.2	3 100	13.1
Propane	2.00	46.4	2 825	3.7
Hydrogen	0.08	120	2 525	8.9

Oxygen

Oxygen is stored as a compressed gas or liquid. In bottles, it is usually stored at a pressure of 20 MPa. Those using oxygen in large quantities usually receive the gas in liquid form.

It is important to make sure that all connections are clean and tight, in order to avoid leakage. If pure oxygen is directed at something flammable, a fire can start very easily. In particular, oil or grease must never be applied to connections.

Pressure regulators

The purpose of the pressure regulator is to reduce the high and variable pressure in the bottle to a suitable working pressure for the welding torch. It keeps the gas flow rate constant throughout the life of the bottle charge, despite any variations in back pressure caused by the heating of the welding torch.

Gas hoses

Gas hoses are colour-coded: red for acetylene and blue for oxygen. In addition, in order to protect against mistakes, the acetylene connection has a left-hand thread, while the oxygen connection has a right-hand thread.

Backfire and flashback

A *backfire* means that the flame burns backwards into the torch with a popping sound. It occurs if the combustion speed of the flame exceeds the speed at which the gas is being supplied, so that the flame front moves backwards. In the case a of *sustained backfire*, which gives a whistling sound, all gas valves must immediately be shut off.

A *flashback* is caused by the reverse flow of gases upstream into the hoses, e.g. by oxygen having entered the acetylene hose and thus forming an explosive mixture. A flashback arrester fitted at the regulator prevents a flashback from going any further back and reaching the acetylene bottle where it would trigger an explosive decomposition.

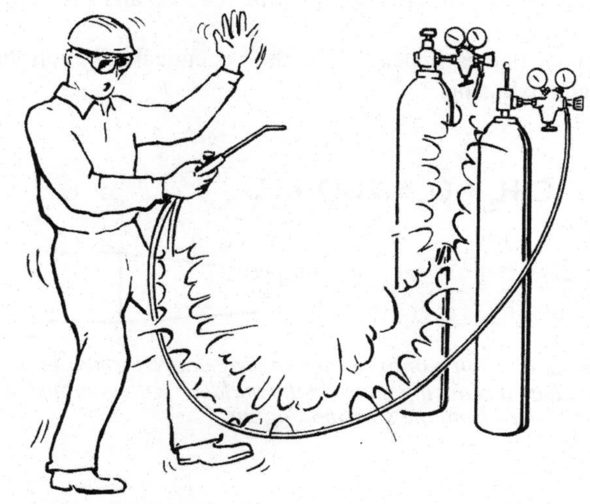

Figure 2.2 Flashback.

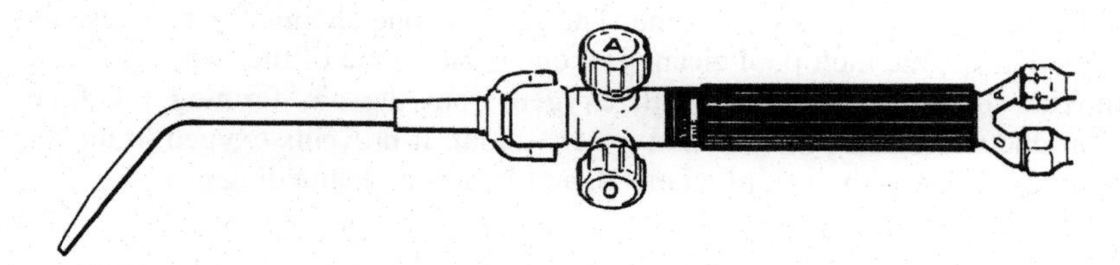

Figure 2.3 Gas welding torch.

Welding torches

A typical welding torch is shown in Figure 2.3. One can distinguish between two types of welding torches: *injector torches* for low pressure acetylene and *medium pressure torches* for high pressure acetylene.

In high pressure torches, the acetylene and oxygen flows are powered by the pressure in their storage bottles, and mix in the mixing chamber section of the torch.

In low-pressure torches, the oxygen flows into the torch through a central jet, producing an injection effect that draws in acetylene from the surrounding connection. From here, the gases continue to the mixing section in the welding torch prior to combustion.

2.3 Gas flames

The basic requirement for a good weld is that the size and type of the flame should be suited to the type of work.

The size of the flame depends on the size of the torch nozzle and on the pressure of the gases flowing through it. This pressure should be maintained within certain limits. If it exceeds the normal pressure, there will be a considerable jet effect and the flame will become 'hard'. Below the correct pressure, the jet effect will be reduced and the flame will be 'soft'.

There are three different types of flames, depending on their chemical effect on the melt pool: neutral, carburising and oxidising.

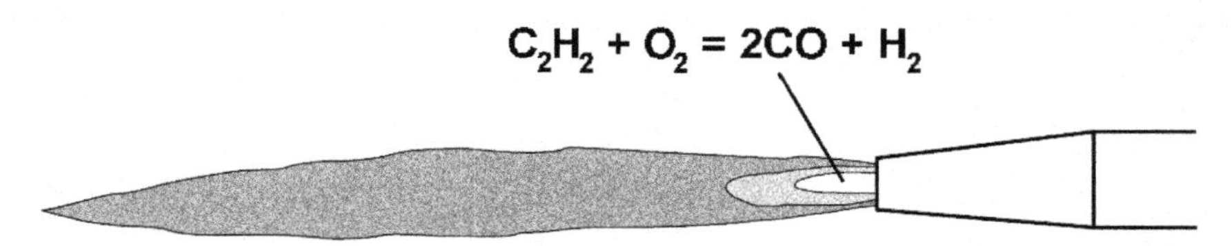

Figure 2.4 A neutral welding flame. Carbon monoxide and hydrogen are formed in the innermost reaction zone. They produce a reducing zone (in the middle), with combustion continuing in the outer zone with oxygen from the surrounding air.

Neutral flame

The neutral flame (see Figure 2.4) is used most. It is easily recognised by the three clearly distinguished combustion zones. The innermost zone, the cone, is a mixing zone and glows white. Acetylene burns here to form carbon monoxide and hydrogen which produce a colourless tongue around the cone. The second zone chemically reduces any metal oxides and keeps the melt pool clean. The outer, blue zone of the flame is where carbon monoxide and hydrogen burn with oxygen from the air, forming the final combustion products of carbon dioxide and water vapour. It prevents oxygen in the air from coming into contact with the molten metal, and so acts as a shielding gas.

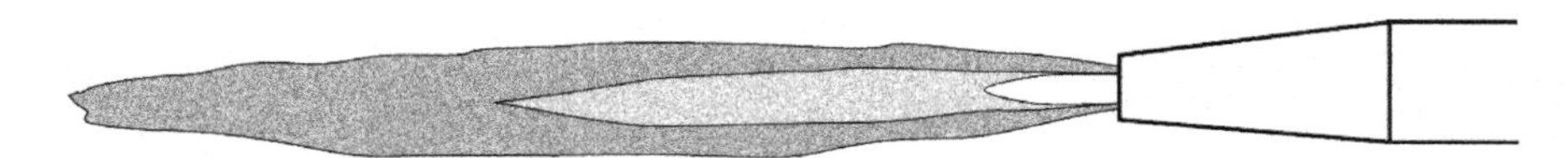

Figure 2.5 Carburising flame.

The carburising flame

If the proportion of acetylene in a neutral flame is increased, there is insufficient oxygen to burn the surplus acetylene in the core zone. The acetylene therefore continues to the second zone, where it appears as a highly luminous yellow-white flame (see Figure 2.5). To some extent, the length of second zone indicates the amount of excess acetylene.

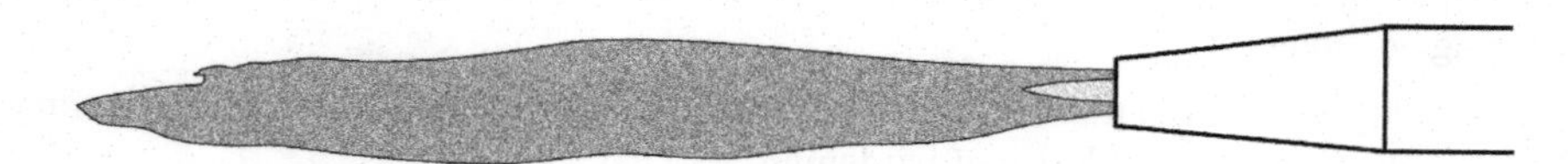

Figure 2.6 Oxidising flame.

The oxidising flame

If the quantity of oxygen is increased, the flame changes to an oxidising flame. The core length is reduced, and the flame takes on a violet tinge with low luminosity (see Figure 2.6).

2.4 Welding techniques

Two different methods of welding are used when gas welding: forehand and backhand. The flame in forehand welding is directed away from the finished weld, while in backhand welding it is directed towards it (Figure 2.7).

Thin sheet metal (less than 3 mm) is normally welded using forehand welding. Steel over 3 mm thick should be backhand welded as it will give the necessary deep penetration and help gases and slag to escape from the large melt pool. Backhand welding is also faster than forehand welding. This means the workpiece is subjected to high temperature for a shorter time. As a result, backhand welded materials have a finer crystalline structure and retain their toughness better than would have been the case if they had been forehand welded.

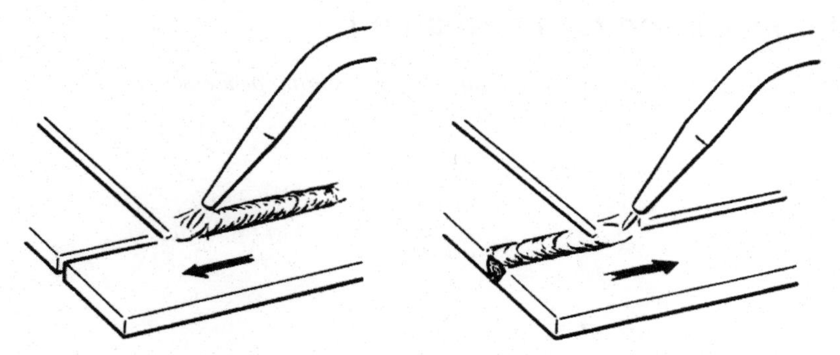

Figure 2.7 Forehand welding (left) and backhand welding (right).

Flux is used when welding easily oxidised materials, where the welding flame itself is insufficient to prevent oxides forming. This is likely to be the case when welding stainless steels and non-ferrous metals. The flux is brushed onto the joint surfaces before welding, and must be thoroughly removed after welding in order to prevent corrosion.

2.5 Applications

Gas welding is very suitable for welding pipes and tubes. It is both effective and economic for applications such as construction and repair of heating, ventilation and air conditioning (HVAC) systems. Other applications of gas welding include welding of hot water pipes, gas bottles, nuclear heat exchangers and boilers. The technique has the following advantages:

- The ability to even out the temperature in the weld at low temperatures. Slow heating and cooling can avoid the risk of hardening.
- Metal thicknesses up to about 6 mm can be welded with an I-joint.
- Speed, as only one pass is needed. Filler wires can be changed without having to pause for grinding.
- Good control of melting, as the welder can see at all times that he has the desired pear-shaped opening in the bottom of the melt pool.
- Root defects are avoided by taking care to ensure good burn-through.
- Pipes and tubes often have to be welded in very confined spaces. In such cases, gas welding is often preferable, bearing in mind the less bulky protective equipment required (goggles, as against a normal arc welding helmet or visor, and compact torch) to perform the work.
- The equipment is easy to transport and requires no electricity supply.
- It is possible to use the light from the flame to locate the joint before welding starts.
- The size of the HAZ can be reduced by surrounding the weld area with damp (fire-proof!) material.

Warning: Note the risk of fire when carrying out temporary welding or cutting work in the vicinity of flammable materials or parts of buildings.

2.6 References and further reading

American Welding Society, *Welding Handbook: Vol. 2, Welding processes.*

3 Basics of electricity in welding

3.1 Introduction

This chapter will give the knowledge needed to understand the basics of electricity, key electronic components used in power sources and techniques to make measurements in welding.

3.2 Basic electrical concepts

Current and voltage

An *electric current* refers to a flow of electrically charged particles. If the number of free electrons in a conductor increases or decreases, the effect is to create an electric charge, or *potential*. As electrons have a negative charge, a surplus of electrons causes a negative potential, and a deficit of them causes a positive potential. It is the difference in potential that is referred to as *voltage*, and which powers the flow of electrons. Pressure and flow in a gas or liquid are good comparisons. Voltage is measured in volts [V], and current in ampères [A], (usually referred to simply as amps).

Power sources

For a current to flow, there needs to be a closed electrical circuit and some means of creating the necessary voltage. Where higher powers are required, the power source is a generator that converts mechanical energy into electrical energy. For small transportable units, it is more practical to use batteries, which convert chemical energy into electrical energy.

For welding applications, we use the name 'power source' for a unit that converts electrical energy from the mains to a form suitable for welding.

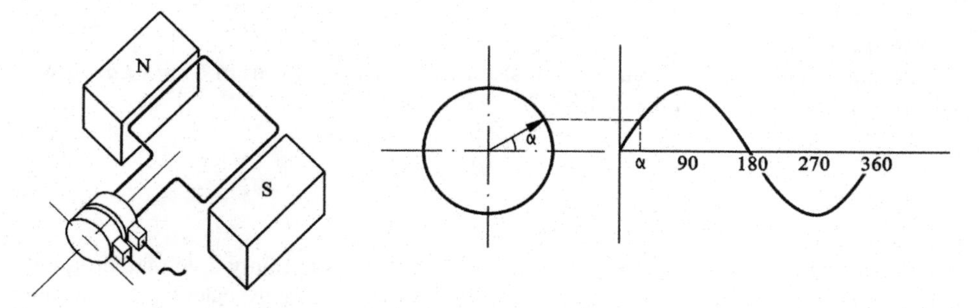

Figure 3.1 The principle of an AC generator (alternator).

Direct current and alternating current

A power source can provide direct current (DC) or alternating current (AC). A battery provides direct current, while a simple generator provides alternating current. The characteristic of an alternating current is that the current changes direction at a certain frequency. Electricity supplies in Europe do so with 50 cycles per second, referred to as 50 hertz [Hz].

The advantage of alternating current is that its voltage can easily be increased or decreased in a *transformer.* When transmitting electricity over long distances, it is advantageous to do so at a high voltage, as the current is correspondingly reduced and so the losses in the power lines are also reduced.

Average value or RMS[1] value

It is important, when measuring the current or voltage with an ammeter or voltmeter, to understand the difference between *average value* and *RMS value*. Pure direct current is not very common, but often includes pulsations, e.g. when welding. In such cases, we use instruments that show the average value of the current.

We use the RMS value when measuring alternating current or alternating voltage. The RMS value of an alternating current is that value of the direct current that would deliver the same power as the alternating current. The following relationship applies for the 230 V alternating voltage that is the usual value of power socket outlets:

Average value[2]	207 V
RMS value	230 V
Peak value	325 V

Resistance

Different conductors have various ability to conduct electricity. A long and thin conductor, such as a wire, has an electrical *resistance* that depends on its length *(L)*, its cross-sectional area *(A)* and the *resistivity* (ρ) of the material. Resistance is measured in ohms [Ω], and can be calculated as follows:

$$R = \rho \cdot L / A$$

Resistivity is temperature-dependent: for most metals, it increases by about 0,4 % per °C.

Ohm's law

The voltage drop (V) of a conductor or resistor can be calculated from *Ohm's Law* if we know its resistance *(R)* and the current *(I)*:

$$V = R \cdot I$$

Power

When a current flows through an electrical resistor, there is not only a voltage drop but also generation of heat. The *power (P)* of this heat can be easily calculated from the formula

$$P = V \cdot I$$

1. RMS: an abbreviation of **R**oot **M**ean **S**quare, a mathematical treatment of the values.
2. To be strictly correct, the mean value of an alternating current is zero, as the positive and negative parts of the cycle are equally large. The average value here is that of the rectified mean voltage.

or

$$P = R \cdot I^2$$

Power is measured in watts [W].

TABLE 3.1 Some electrical quantities and units.

Quantity		**Unit**	
Name	**Indicated by**	**Name**	**Indicated by**
Current	I	ampère	A
Voltage	V	volt	V
Power	P	watt	W
Energy	W	joule	J
Frequency	f	hertz	Hz
Resistance	R	ohm	Ω
Inductance	L	henry	H
Capacitance	C	farad	F
Magnetic flux	Φ	weber	Wb
Magnetic flux density	B	tesla	T

TABLE 3.2 Common prefixes for powers of ten.

n (nano)	10^{-9}	k (kilo)	10^3
μ (micro)	10^{-6}	M (mega)	10^6
m (milli)	10^{-3}	G (giga)	10^9

Electrical circuits

The combined resistance of two resistors connected in *series connection* is simply the arithmetical sum of the resistances of the two resistors. The same current flows through both of them.

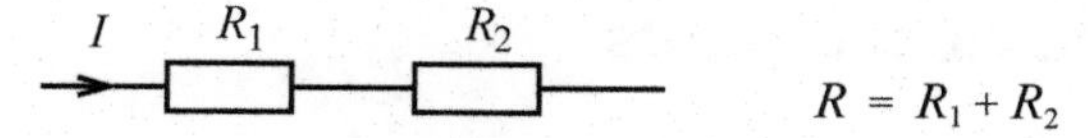

$$R = R_1 + R_2$$

Calculating the resistance of two resistors connected in *parallel connection* is a little more complicated:

$$\frac{1}{R} = \frac{1}{R_1} + \frac{1}{R_2} \quad \text{or} \quad R = \frac{R_1 \cdot R_2}{R_1 + R_2}$$

The voltage across each of the resistors is the same.

Kirchhoff's laws

Kirchhoff's laws are helpful when analysing an electric circuit with several components in order to work out the current through, or voltage across, any of the components.

Kirchhoff's current law:

"The sum of the currents at a node = 0". This is valid as long as we allow for the directions of the current. Alternatively, it can be simpler to express it as saying that the sum of the currents flowing into a node is equal to the sum of currents flowing out of it.

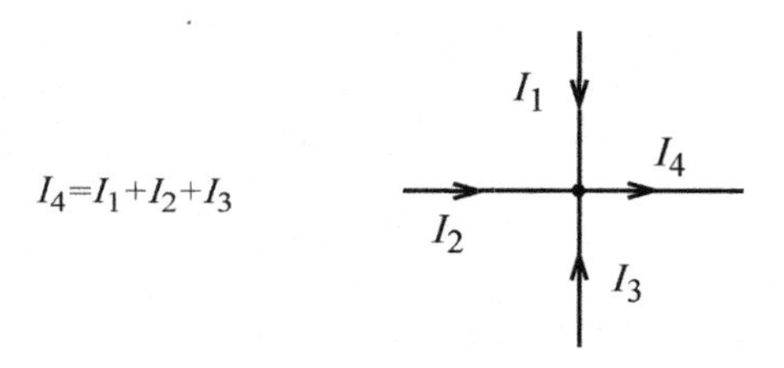

Kirchhoff's voltage law:

"The sum of the voltages around a closed circuit = 0". Here, too, we must think of the signs (positive or negative). If we follow a closed loop in a given direction, the sum of all the voltages in that direction equals the sum of all the voltages in the opposite direction.

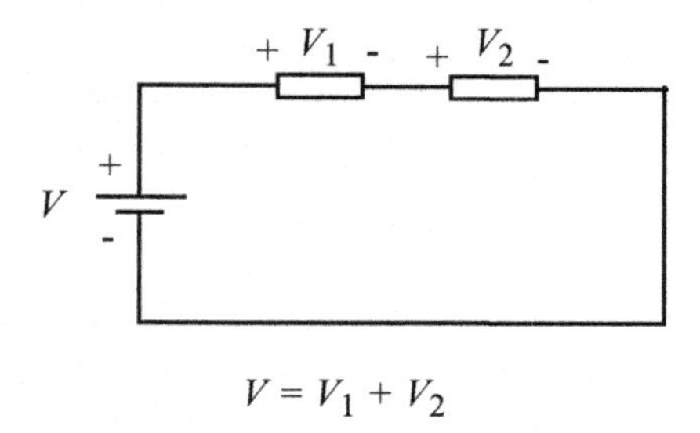

3.3 Components in electrical circuits

The capacitor

A capacitor consists in principle of two metal plates with insulation between them. If the area of the plates is very large, and the insulation between them is very thin, the capacitor can be given an electric charge. The higher the charge, the higher the voltage between the plates.

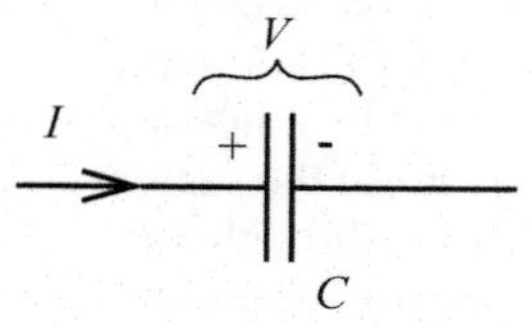

Figure 3.2 Symbol for a capacitor, showing quantities and units.

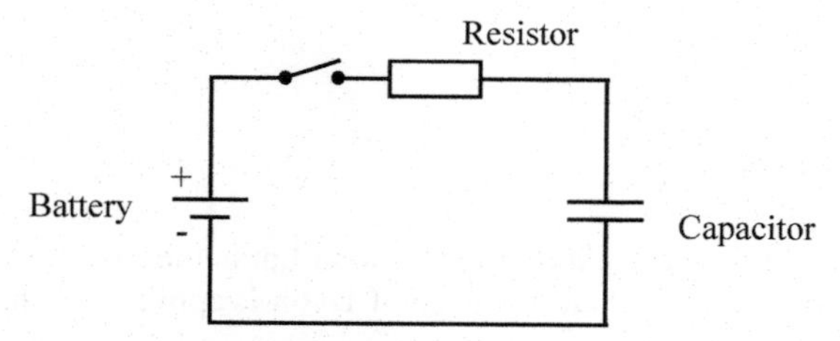

Figure 3.3 Charging a capacitor.

The ability of a capacitor to hold charge is referred to as its *capacitance* (*C*), and is measured in *farads* [F]. The farad is a very large unit: as it is difficult to make capacitors with high capacitances, the most commonly encountered measure is the *microfarad*, [μF]. A common use for capacitors is to smooth a direct voltage which contains an alternating voltage element (ripple), by means of a smoothing capacitor. These capacitors are often of a particular type, known as electrolytic capacitors, which can have high capacitances and which can withstand high currents. However, they can withstand only direct voltage, not alternating voltage, and it is important that they are connected to the correct polarity, i.e. to the correct plus and minus terminals.

A high-capacitance capacitor has little resistance to alternating current, but will not pass direct current. Raising the frequency of the alternating current allows more current to pass, without increasing the voltage drop across the capacitor.

Note that although a capacitor appears to present resistance to an alternating current, and causes a volt drop across itself, no heat is produced as is the case with a resistor. A capacitor merely charges up with, and discharges, energy.

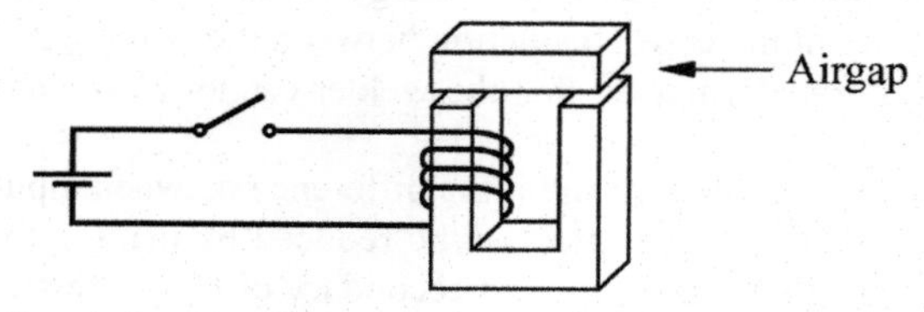

Figure 3.4 An inductor.

The inductor

An inductor (or choke) consists of a coil, i.e. a certain number of turns of winding wire. To obtain a sufficiently high value of inductance, this coil is wound around an iron core. A magnetic material, iron, conducts magnetic flux much better than does air. However, there is often a small air gap in the iron core, which prevents the magnetic flux from exceeding the maximum that the iron can carry. If the flux density in the iron core is too high, the iron will become saturated and unable to carry any higher magnetic flux. If the current is increased beyond the saturation value, the inductance falls off rapidly.

We can compare the effect of an inductor with a flywheel or a heavy vehicle, for either of which a high force is needed in order to accelerate it. In the same way, a high voltage is needed in order quickly to change the current flowing through a coil.
If, for example, we open a switch in a circuit with an inductor when the inductor is carrying current, there will be a brief arc (an electric spark), which can damage the switch contacts.

I

L

Figure 3.5 Schematic symbol for inductance.

Inductance (L) is measured in henries [H]. An inductor which limits the rate of rise of current through it to 1 ampere per second when a voltage of 1 volt is applied to it has an inductance of 1 henry.

In many respects, the inductor is the opposite to the capacitor. The voltage across a capacitor increases gradually as current flows into it, while the current flowing through an inductor increases gradually when a voltage is applied to it. Inductors also resist the flow of alternating current. The voltage drop across an inductor increases as the frequency increases, which is the exact opposite of the behaviour of a capacitor. An inductor passes DC with no voltage drop other than that caused by the resistance of the windings.

The main function of the inductor in a MIG welding rectifier is to reduce current peaks at the drop short circuits.

The transformer

A transformer converts (transforms) one alternating voltage to another. This means that, for example, a welding power source can convert a 400 V mains supply to 40 V, as needed for welding, and which is also safe for the welder.

A transformer consists of an iron core with two windings: a *primary winding* for the incoming power supply, and a *secondary winding* for the outgoing power. The two windings are normally electrically insulated from each other: energy is transferred to the secondary winding by the magnetic flux in the core. Insulation between the windings is very important for safety in a welding transformer, so that the welder cannot come into contact with mains voltage.

If there is no requirement for electrical insulation, and if the difference between input and output voltages is small, the cost of the transformer can be reduced by using only one winding, and making it serve for both the primary and secondary circuits: this is known as an *autotransformer.*

A three-phase transformer has three legs, each having primary and secondary windings around it.

It is desirable that, when the secondary winding is not connected to any load, the current in the primary winding should be low. This is done by giving it a sufficiently large number of turns to produce a high inductance.

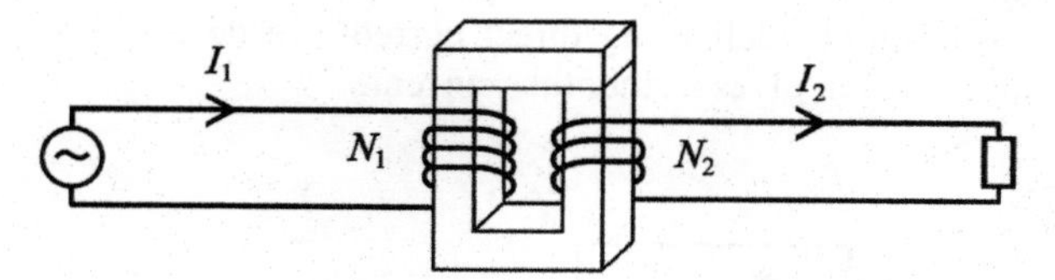

Figure 3.6 A transformer.

As the magnetic flux through the primary and secondary windings is the same, there is the same voltage per turn in the two windings. The ratio between the number of turns in the two windings (N_1 and N_2) is referred to as the *turns ratio*, and therefore also determines the ratio between the primary and secondary voltages.

$$\text{Turns ratio} = \frac{N_1}{N_2} = \frac{V_1}{V_2}$$

This means that, by varying the number of turns in the secondary winding, the output voltage can be adjusted to suit the application.

The current through the windings is also dependent on the turns ratio. If the primary winding has more turns than the secondary winding, the current through it will be lower.

$$I_1 = \frac{N_2}{N_1} \cdot I_2$$

The rectifier

A rectifier converts alternating current to direct current, using *diodes*. A diode can conduct current in only one direction.

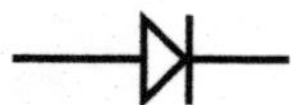

Figure 3.7 Schematic symbol for a diode.

Four diodes form a single-phase rectifier bridge.

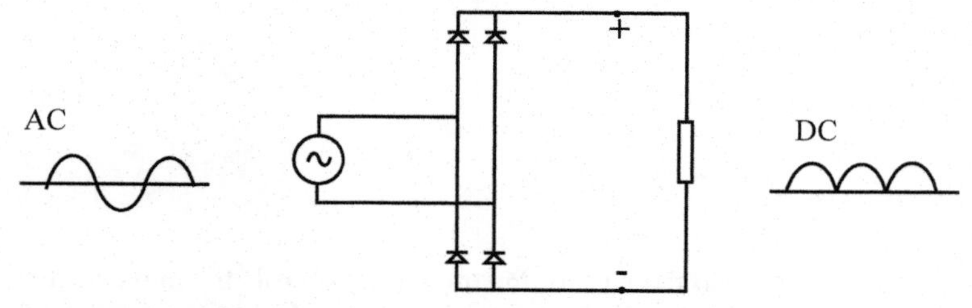

Figure 3.8 Four diodes form a single-phase rectifier bridge.

A single-phase rectifier bridge, consisting of four diodes, produces a pulsating direct current. Almost all welding power sources delivering direct current are designed for operation from a three-phase supply, which gives a smoother current.

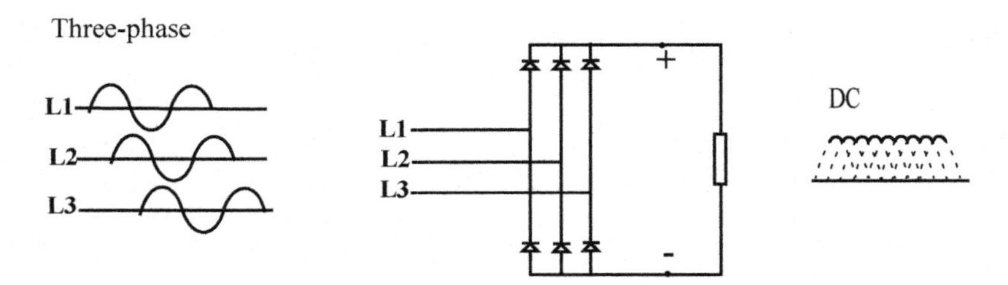

Figure 3.9 Three-phase rectifier bridge.

Using *thyristors* produces a controllable rectifier, i.e. one in which the output voltage can be steplessly regulated. Thyristors, which have a third terminal (the *gate*) in addition to the input and output power connections, can be used in a rectifier circuit instead of diodes. The thyristor does not conduct until it receives an ignition pulse via the gate connection. It continues to conduct until the current falls to zero, and must then receive another ignition pulse when it is due to conduct again. The gate pulses are provided by an electronic control circuit.

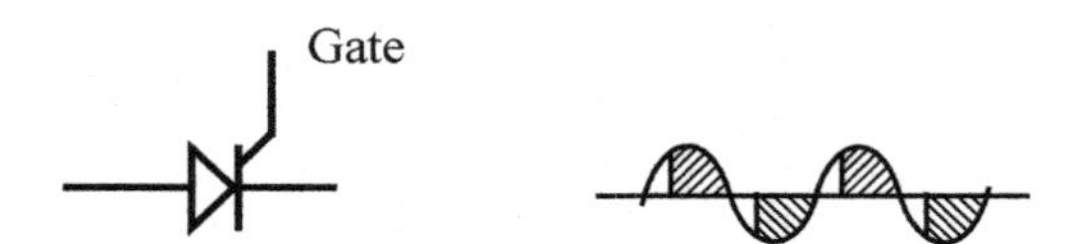

Figure 3.10 The thyristor is triggered with a controlled delay after each zero crossing, which therefore limits the output voltage to the desired value.

The transistor

Transistors can also be used to control the output current and voltage in power circuits.

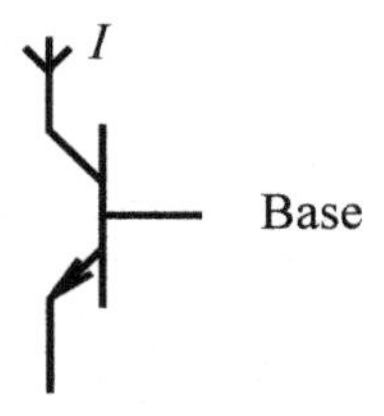

Figure 3.11 A transistor.

A small control current into the base of the transistor controls the main current flow (I) through the transistor. This means that the transistor can be used as an amplifier, as the main current through it is a more powerful copy of the control current.

In exactly the same way as in a resistor, there are power losses that are released as heat, and which must be removed by cooling to prevent the transistor from overheating. These losses can be reduced by using the transistor instead as a switch, using a high frequency control signal to deliver the output current in pulses. The current can be controlled by, for example, varying the pulse duration *(pulse width modulation)*.

Transistors come in many types, depending on how they are made. The type described above is a *bipolar junction* transistor. What is known as the *field-effect transistor* (FET) is controlled by a control voltage rather than by a control current. Another abbreviation is MOS (Metal-Oxide Semiconductor), which indicates the method of construction. High-power MOS-FET transistors (or, as they are often known, Power-MOS transistors) are very suitable for controlling the welding current in a switched power source. They need only little power for control, and can carry high currents and very rapid pulses.

The *IGBT transistor* (Integrated-Gate Bipolar Transistor) is a combination of a bipolar transistor and an FET transistor. It needs only little power for control, and withstands high voltages better than Power-MOS transistors, but is not as fast.

3.4 Measuring welding data

Measuring current and voltage

Welding voltage is often measured at the power source terminals, e.g. by an instrument built into the power source: see Figure 3.12. However, this introduces a source of error from the voltage drop across the welding cables. For correct measurement of the actual welding voltage, it is therefore important that the voltage should be measured as close to the welding position as possible.

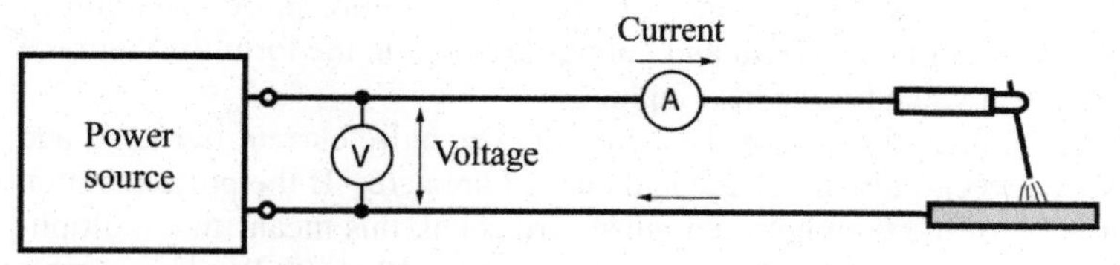

Figure 3.12 The welding circuit.

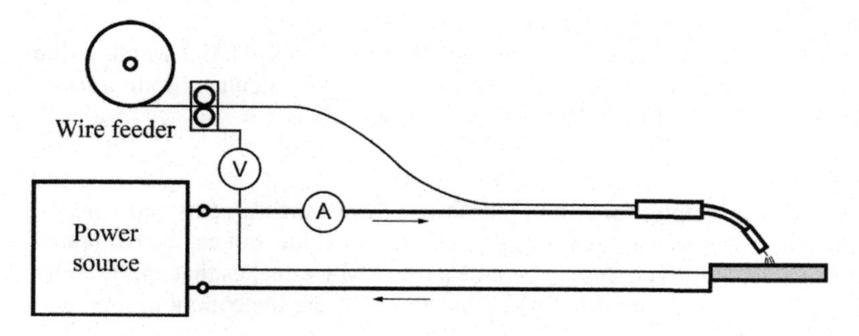

Figure 3.13 Measuring the welding voltage between the wire feeder and the workpiece avoids the volt drop of the welding current cables.

In MIG/MAG welding, for example, we can measure the voltage between the wire at the feed rollers and the workpiece, as is shown in Figure 3.13. The wire itself serves as an ideal conductor right up to the arc, thus avoiding the voltage drop over the welding cables. The wire and the wire feeder must, of course, be electrically insulated from the welding power source right up to the welding gun, as shown in the diagram.

The *welding current* is the same anywhere in the welding current circuit, so it does not matter where it is measured in the circuit. However, it is usually so high that it cannot be passed directly through the instrument: some other method is needed. There are various options:

- We can measure the volt drop across a *shunt*, which is a resistor with a small but carefully calibrated resistance, fitted somewhere in the welding current circuit. If we want to measure rapidly changing currents, it is important that the shunt has as low inductance as possible, as otherwise this would provide a spurious addition to the measured value.
- Instead of a shunt, there are also sensors based on measurement of the magnetic field around a conductor, known as *Hall-effect elements*.
- *Clip-on ammeters* are simple to use for occasional measurements.
- *Current transformers* are used to measure high alternating currents. These are transformers in which the core surrounds the welding current conductor, i.e. so that the conductor is the single turn of the 'primary winding'. The secondary winding is a coil with several turns, which means that the current in it is therefore reduced in proportion to the turns ratio (see above) to give a value suitable for the measuring instrument.

Calculating the heat input at pulsed welding

Calculation of heat input as described on Page 9 does not apply for pulsed welding. If the measured average values of the current and voltage are used in the formula, there is a risk of obtaining too low a value for the heat input.

Example: A welding operation uses pulsed current. The pulse current is 100 A, and the voltage is 20 V: between pulses, voltage and current are zero. If the pulse duration (width) and the duration of the time between pulses are equal, this means that multiplication of the average values would give a power of 50 x 10 = 500 W. The correct method is to calculate the instantaneous power during a pulse (100 x 20 = 2000 W), and then to multiply by the pulse factor, which is 0.5 as the pulse width and pause width are the same. The true power is therefore 2000 x 0.5 = 1000 W.

If the current is pulsed between two values, such as with pulsed MIG welding, the heat input of the pulse and of the background current must be calculated individually, and then be added in proportion to their respective time durations.

Instruments

There are two working principles for ordinary analogue display instruments: *moving coil* or *moving iron*. Moving coil instruments measure the mean value, but can be calibrated to show the RMS value. However, this will not be a true RMS value, as that applies only for a sinusoidal alternating current. Moving coil instruments are the commonest type of analogue display instruments.

The measurement principle of moving iron instruments is such that they directly measure and display the RMS value.

Digital instruments can generally be designed so that they display the required measured result, i.e. the average value of the DC component and the RMS value of the AC component. However, it is not necessarily the case that they measure the true RMS value of AC voltage.

When used for measuring welding data, digital instruments should be designed so that the displayed value is not constantly altering. If the instrument is supposed to provide an indication of the stability of welding, it can be more suitable to use an instrument having an analogue display.

Measuring other welding parameters

Wire feed speed

Special sensors are available, having idler rollers that generate a signal proportional to the wire feed speed. For occasional check measurements, it may be sufficient to use a tape to measure the length of wire delivered by the feeder in a certain length of time, from which the speed can be calculated.

Shielding gas flow

The commonest method of measuring the gas flow is to use a marked tapered glass tube with a float inside. When using this to check performance, it should be held vertically against the welding gun nozzle. Sensors that provide an electrical output signal are also available.

Temperature

It is important, when measuring the temperature of the workpiece, to measure in direct thermal contact with the workpiece. One simple method is the use of special *chalks* that change colour to indicate particular temperatures.

A contact thermometer is an instrument with a measurement tip that is pressed against the workpiece.

A *thermocouple* is appropriate for use when continuously monitoring a temperature. It consists of two wires of dissimilar metals that are soldered together at one end: the contact between the two metals produces a thermoelectric voltage that is proportional to the temperature. If the other end of the wires is hold at a known temperature, e.g. 0 degrees, the voltage reading is directly related to the temperature of the soldered junction. It is important that the soldered tip is in good thermal contact with the part to be measured, e.g. by placing it in a drilled hole.

A *pyrometer* is an instrument that measures temperature remotely, i.e. by measuring radiation from the object, without having to be in contact. Pyrometers are also available with electrical signal outputs for recording or other purposes.

If we need to see the temperature distribution over an area, this can be done using a thermal imaging *IR camera*, which displays the temperatures in the form of different colours on a screen.

4 Arc welding: an overview

4.1 Introduction

In a typical arc welding process there are many important parameters that influence quality and the production rate. Knowledge of arc physics, the forces acting on the arc and the interaction of the arc with the different shielding gases is essential in order to understand how to optimise the welding process.

4.2 Arc physics

A welding arc is an electrical discharge between two electrodes. The welding current is conducted from the electrode to the workpiece through a heated and ionised gas, called plasma. The voltage drop and current in the arc determine the amount of electric power that is released, the heat from which melts the electrode and the joint faces, allowing welding to take place.

The power must be high enough to keep the temperature of the arc sufficient for the continued transport of the current. The temperature maintains the ionisation of the gas or plasma, i.e. it creates electrically charged particles that carry the current. Ionisation occurs as a result of collisions between the particles in the plasma being so powerful that free electrons are emitted. The resulting electron-deficient atom or molecule is known as an ion, and is positively charged as a result of losing one or more electrons. The charged particles are accelerated by the welding voltage and transfer their energy by means of multiple collisions.

Depending on the choice of shielding gas, different temperatures are needed to keep the plasma ionised. Argon, for example, is easier to ionise than helium. Welding in helium or helium-mixed gases produces a higher voltage drop and higher heat input to the weld pool.

If the arc has been struck in a multi-atomic gas, such as CO_2, the molecules will first dissociate (i.e. be divided up into atoms) as the temperature rises. Dissociation requires energy, which is taken from the hotter parts of the arc, and lost when the atoms or molecules recombine at lower temperatures, which means that the thermal conductivity of the arc increases.

Voltage distribution and heat generation

An arc can be divided into three regions from the point of view of voltage distribution (Figure 4.1):

- the anode region
- the arc column
- the cathode region.

The voltage drops across the anode and cathode regions are constant, i.e. they are relatively independent of the welding current and the welding voltage. Together they can amount to more than half of the total arc voltage, despite occupying only a negligible part of the total length of the arc.

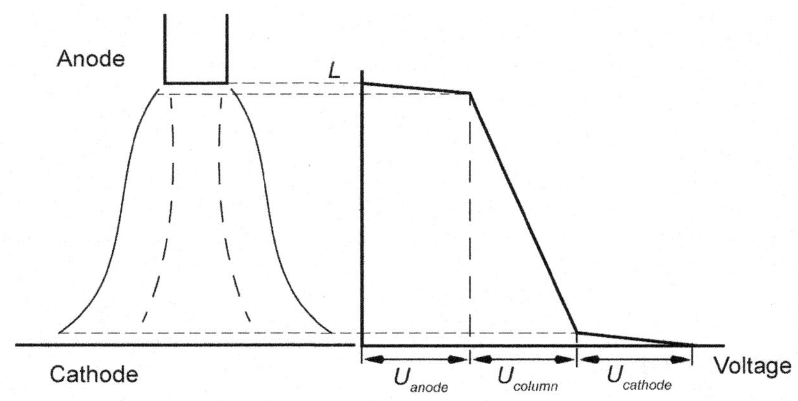

Figure 4.1 Voltage distribution in the arc.

Anode region

The closeness of this part of the arc to the anode means that a lot of heat is conducted away. As the temperature falls, so a higher voltage is required in order to maintain ionisation in the arc. The heat loss is compensated for mainly by the free electrons in the plasma having to accelerate in a sufficiently strong field between each collision, which requires a high voltage drop.

TABLE 4.1 Work function for a number of metals.

Metal	Work function, eV
Iron	4.79
Copper	4.82
Aluminium	3.95
Barium	2.29
Caesium	1.36
Tungsten	5.36
Thorium	3.57
Tungsten with thorium oxide	2.62

Cathode region

Here, too, the volt drop depends on the amount of heat conducted away. It might therefore be thought that the same amount of heat is delivered to the anode and the cathode regions. That this is not so is due to the electrons that must be liberated from the cathode, which requires an additional voltage known as the work function. This energy is recovered as heat at the anode when the electrons recombine with the metal. As a result, more heat is generated at the anode than at the cathode. The work function varies from one metal to another, as shown in Table 4.1.

Thermal and field emission

Electron emission occurs as the result of a combination of two principles: thermal emission and field emission. In thermal emission electrons are emitted from the cathode as a result of high temperature. The current density that can be delivered depends largely on the temperature. Most metals have a boiling point that is so low that it sets a limit to the permissible temperature. Tungsten, however, can operate at a temperature of over 3000 °C, giving good thermal emission and a stable arc. When using consumable welding electrodes, thermal emission from the metal is not sufficient. A strong electric field is also needed in order to release the electrons.

The current from the cathode is not uniformly distributed across the metal, but is concentrated in one or several small mobile cathode spots with very high current density. The spots wander around randomly, removing oxides from the cathode surface. At the same time oxides assist electron emission and means the work function energy is lower. TIG electrodes, for example, perform better if small quantities of certain metal oxides are added to them. In MIG welding, the shielding gas can improve arc stability if an oxidising gas is added to the shielding gas. Pure argon gives relatively poor arc stability, but even small quantities of O_2 or CO_2 help stable cathode spots to form on the surface of the weld pool (the cathode), and the arc to burn stably, which produces a smooth weld bead.

The arc column

The voltage drop in the actual arc (U_{column} in Fig. 4.1) depends on the length of the arc (about 3–10 mm) and, in particular, on the composition of the shielding gas. Argon plasma has a volt drop of about 6 V/cm, while helium plasma has a volt drop of 42 V/cm. Helium also gives a higher temperature in the plasma due to its higher ionisation energy.

TABLE 4.2 Ionisation energy of some gases and metals.

Metals	eV	Gases	eV
Cs	3.9	H	13.5
K	4.4	O	13.6
Na	5.1	CO_2	14.4
Al	6.0	CO	14.1
Cr	6.8	N	14.5
Ni	7.6	Ar	15.7
Fe	7.8	He	24.5

When welding with consumable electrodes, there is substantial evaporation of metal from the molten drop at the tip of the electrode. Metal vapour has a low ionisation potential, and is responsible for a significant proportion of the current transport. Processes protected by molten flux, e.g. the use of coated electrodes, – also include substances in the flux that stabilise the arc.

The temperature of the arc, in other words, is affected by the substances that are present. High concentrations of easily ionised metals, such as sodium or potassium, help the arc to burn stably at temperatures around 6000 K. In the presence of difficult-to-ionise inert gases, the arc temperature can be as high as 30 000 K. Table 4.2 shows the ionisation energies of a number of gases and metals.

The arc voltage drop is not as dependent on the current as might be expected from Ohm's Law (see Figure 4.2). This is due to the fact that the anode and cathode voltage drops are constant, as well as to the fact that the number of charge carriers increases and the diameter of the arc increases with higher current. Conversely, at low current, the diameter of the arc decreases, which makes it more difficult for the arc to maintain the temperature, and so a higher voltage is needed in order to maintain sufficient ionisation.

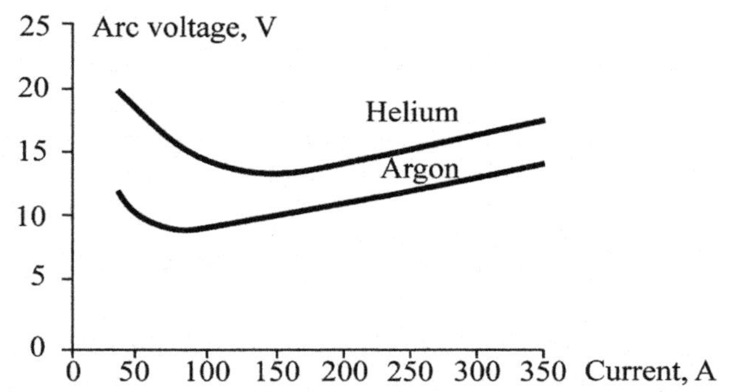

Figure 4.2 Arc voltage drop for TIG welding of stainless steel with an arc length of 3.2 mm (AGA).

Deposition rate

Heat is created where there is a voltage drop, as calculated by the formula $P = U \cdot I$. If we multiply the anode and cathode voltage drops by the welding current, we obtain values for the heat released at the electrode and at the workpiece. The latent heat of fusion (*Ps*) necessary to melt the metal of the electrode is provided mainly by the heat evolved in the anode region of the arc, given by:

$$P_S = k \cdot U_a \cdot I$$

where U_a = the anode volt drop and I = the welding current. Constant k is an efficiency factor that indicates what proportion of the heat is used for melting the electrode. Some of this heat is used as sensible heat in the molten drop (see Figure 4.3).

As the anode voltage can be considered as a constant voltage drop, heating power is proportional to the current. The melting rate is therefore also proportional to current and, at least in theory, independent of the voltage setting.

Penetration

The depth of melt penetration into the workpiece depends on several factors. Heat supply is of course important, but so is the energy density. If energy density is low, a larger surface must be heated, which means that the cooling area is high. At higher current densities, the pressure of the arc creates a crater that improves penetration. At very high energy densities, the metal on the surface of the weld pool is evaporated so strongly that it creates a reaction force that increases penetration. In the extreme case, the crater becomes very deep and narrow, forming what is known as a keyhole.

If the electrode is held other than perpendicular to the workpiece, penetration will be altered by the arc either pushing the molten metal away or pushing it in front of the arc, depending on the angle of the electrode.

Several other factors also affect the penetration. Currents (i.e. convection currents, not electric currents) in the weld pool help to carry down heat from the arc to the unmelted base material. They can be caused by electromagnetic forces, the surface tension phenomenon (the Marangoni effect) or by thermal currents (the Buoyancy effect).

Polarity

The polarity of the electrode also plays a part, as there is greater heat evolution at the anode than at the cathode. This is due to the fact that the input of energy to release electrons at the cathode is recovered at the anode in the form of heat as the electrons are absorbed by the anode. It is therefore standard practice, when performing TIG welding, that the tungsten electrode should be connected to the negative pole of the power source. This will mean that there will be less heat load on the electrode, with the heat being concentrated on the workpiece.

When performing (for example) MIG welding using a consumable wire, the wire is normally connected to the positive pole. Connecting it to the negative pole could mean that the effect on the molten droplet at the tip of the wire of the concentrated and randomly moving cathode spots would be to make the arc jittery.

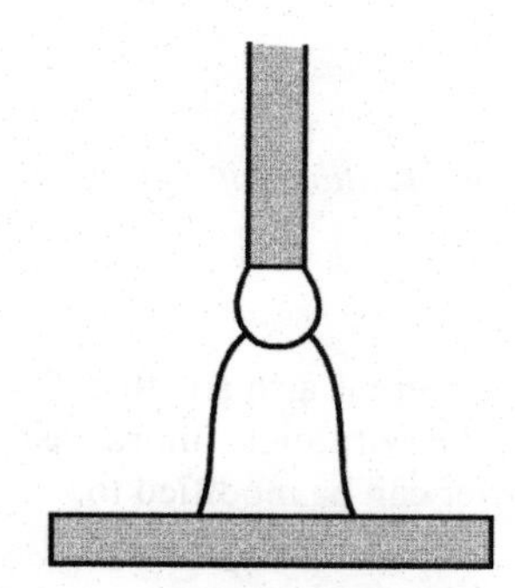

Figure 4.3 The droplet partly prevents the heat from the arc from melting the electrode.

As the heat evolution is greatest at the positive pole, it might be thought that a positive electrode would melt the most quickly. Generally, however, the effect is the opposite. This is because much of the heat developed at the anode superheats the droplets instead of melting new electrode material. The combination of hot droplets and a stable arc means that penetration is good. With a negative electrode, the cathode spots move towards the unmelted parts of the electrode, increasing the efficiency of melting the material of the electrode.

The effect of this is that more, but colder, electrode material is transferred to the weld pool when welding with the negative pole. With a negative electrode, the droplet process is also more jittery.

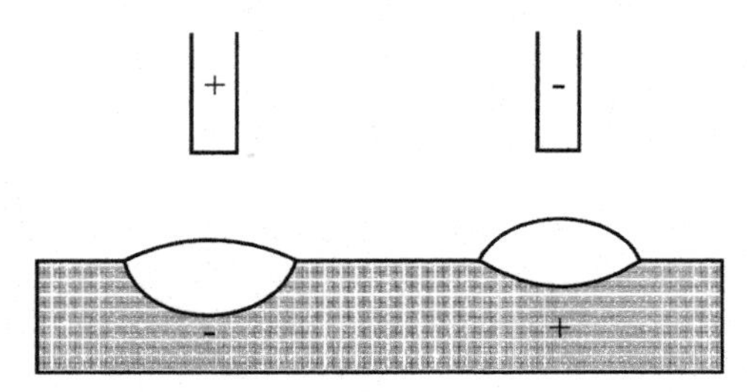

Figure 4.4 When using a consumable electrode, a positive electrode gives good penetration, but has lower deposition rate than a negative electrode.

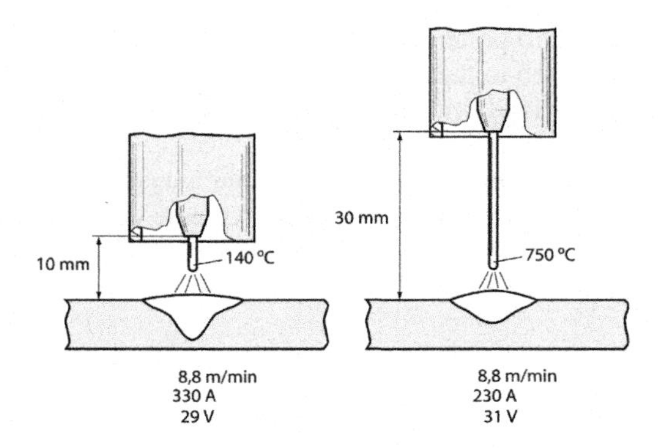

Figure 4.5 The wire temperature rises and the current falls as the stick-out increases.

The effect of wire stick-out length

Depending on the stick-out length and the current in the wire, there is also an effect from resistive heating of the wire. Due to this resistive heating, the wire melts more easily, causing the current to be reduced. The formula for melt power can be modified to:

$$P_S = k \cdot U_a \cdot I + R_e \cdot I^2$$

where R_e = the resistance of the stick-out portion of the wire. As the resistive heating increases, so the current falls in order to maintain the melting rate (see Figure 4.5).

The heat input to the workpiece from the arc falls as the current falls, but the quantity of transferred material remains the same. An unintentional increase of the stick-out length will reduce heat input to the weld and increase the risk of weld defects.

Generation of welding fumes

As a result of the intense heat from the arc, some of the electrode metal is vaporised. When this vapour leaves the arc it is oxidised and forms welding fumes. If some other material is present, having a lower vaporisation temperature, fume generation will increase. Examples of such substances are the flux in a flux-cored wire, or oil, paint or zinc coating on the workpiece.

4.3 Drop transfer

The release of molten droplets from the wire is due mainly to electro-magnetic forces and surface tension. Other forces acting on the droplets are gravity and the drag force from the plasma jet. Gravity is the main problem when holding the weld pool in place during positional welding or single-sided welding of thicker plates without backing material.

Electromagnetic forces and the plasma jet

The flow of current in a conductor creates a magnetic field around the conductor. The current and the magnetic field interact with each other to produce a force acting radially inwards on the conductor. If the conductor is not cylindrical, but for some reason increases or decreases in diameter, the force will have an axial component. This component can accelerate the droplets and create the plasma jet but also causes convection flows of liquid metal in the droplets or in the weld pool.

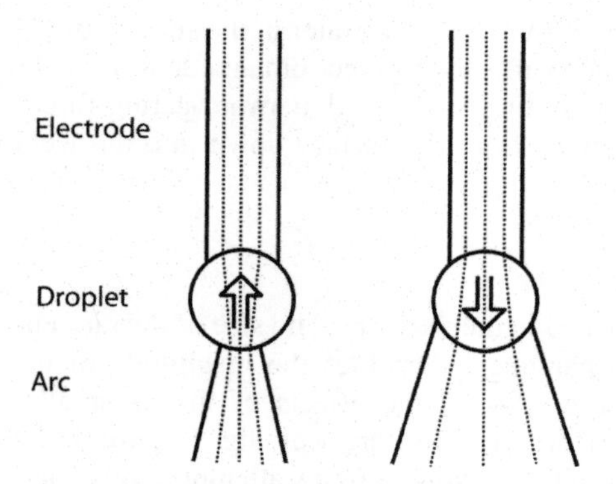

Figure 4.6 The way in which the electromagnetic forces affect the droplet depends on the width of the arc in relation to the electrode diameter. (Note: the diagram is highly schematic.)

If the arc is narrow, with a small anode area, the current is forced to concentrate through a small area on its way out through the droplet, which means that it will be acted on by upwards-acting forces (see Figure 4.6). This is the case with low welding currents or when using CO_2 as the shielding gas. The droplets can then grow and become large (see Figure 4.7). The opposite applies when using a thin wire and an argon-rich shielding gas with a high welding current and voltage. Under these conditions, the anode area is wider and the current paths expand downwards. The forces acting on the arc are now downwards and help to transfer molten electrode material as a stream of small droplets, producing a stable and short circuit-free spray arc.

Electromagnetic forces also affect the plasma column by creating convection currents in it. As the electrode is thin and the workpiece is much wider, the arc becomes bell-shaped. The current density close to the electrode tip is high, which produces a higher pressure there than lower down. This difference in pressure drives a flow of plasma gas from the electrode tip down towards the workpiece, while at the same time new gas is drawn into the upper part of the arc from outside. At higher current densities, the effect is to create a jet stream that can reach a velocity of hundreds of metres per second.

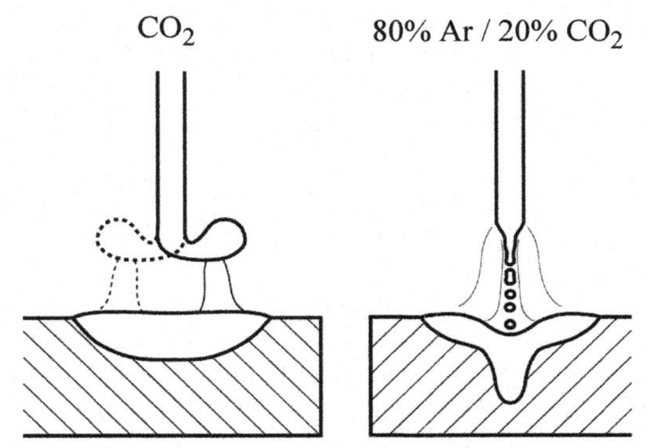

Figure 4.7 The impact from magnet forces on drop transfer. To the left repelled transfer.

The plasma flow produces an "arc pressure" that creates a crater in the surface of the weld pool. Penetration into the workpiece is also affected by electromagnetic forces. As the current paths in the weld pool are divergent, they produce downward-acting forces that produce flows of heated molten metal away from the surface down into the weld pool.

Surface tension

Surface tension acts to keep the droplets attached to the wire tip. The size of the circumference of a wire from which a droplet is hanging determines the magnitude of the resulting upward-acting surface tension force. One of the effects of this is smaller droplet transfer from a thin wire. When performing short-arc welding, (Figure 4.8 f) surface tension helps to transfer the droplets to the weld pool by its attempt to minimise the surface area. Constituents of the shielding gas (e.g. oxygen) and the wire (e.g. silicon) that affect the viscosity and surface tension of the molten metal have a considerable effect on welding behaviour, particularly in short-arc welding.

In TIG welding and some other welding processes, convection currents can arise in the weld pool due to differences in the surface tension at the surface of the pool, known as the Marangoni effect. Depending on their direction and on the presence of surface tension-modifying substances, these convection currents can increase or decrease the penetration of the weld.

Types of drop transfer

When welding with a consumable electrode, such as MIG/MAG welding, the arc has two main functions. One is the above-mentioned supply of heat for melting the materials, and the other is the transport of the molten wire material down to the weld pool.

Material transport is the result of a complicated interaction between different forces. Types of droplet transfer are shown in Figure 4.8. The behaviour of the droplets is mainly dependent on the current and voltage settings and on the composition of the shielding gas. In addition, there are many other parameters that affect transfer to a greater or lesser extent. The most sensitive process is that of welding with short-circuiting material transfer, such as short arc welding. In this case, the dynamic interaction with the power source is extremely important in determining the welding properties.

With a low welding current, the electromagnetic forces are negligible. The molten droplet at the end of the electrode is affected mainly by gravity, and by surface tension which balances the downward gravity force as long as the droplet remains on the electrode tip. As the electrode continues to melt, more material is added to the droplet and so it increases in size. Depending on the arc length, two things can now happen. If the arc is long, the droplet will continue to grow until its weight exceeds the surface tension force, and it then drops. This is known as globular droplet transfer: see Figure 4.8 a.

On the other hand, if the arc length is short, the process is known as *short-arc welding* (see Figure 4.8 f). The droplet comes into contact with the weld pool long before it has become sufficiently large to fall under its own weight. The resulting short-circuit link is interrupted when the droplet transfers. Final nip-off is assisted by magnetic forces when the current density increases in the neck of the droplet: this is known as the pinch effect. The current density increases as a waist forms in the short-circuiting link of molten metal between the electrode and the weld pool, and also as a result of the current from the power source increasing when the arc is interrupted by a short circuit. If all the parameters are properly set, the frequency of short circuits will be high and steady, in the range 50 – 200 Hz.

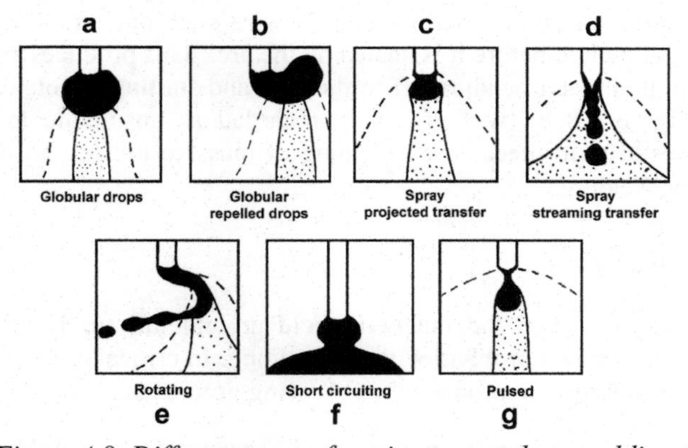

Figure 4.8 Different types of arc in gas metal arc welding (AGA).

When we increase the current to a higher value, electromagnetic forces will dominate over gravity. Again, there are two different cases. If the anode area on the underside of the droplet is small in relation to the cross-sectional area of the wire, the forces acting on the droplet will act upwards and be sufficiently strong to counteract gravity. The droplet will continue to grow, and become large. Due to its instability, it will tend to swing to the side and, when it is finally nipped off, it will receive a kick from the electromagnetic forces and will fly off with a rotary motion diagonally down towards the workpiece (see Figure 4.8 b). This is known as repelled transfer and is typically a problem in CO_2-shielded welding. One way of reducing this unacceptable form of droplet transfer is to reduce the voltage and change to short-arc welding. However, short-arc welding at high currents produces considerable spatter, particularly if CO_2 is used as the shielding gas.

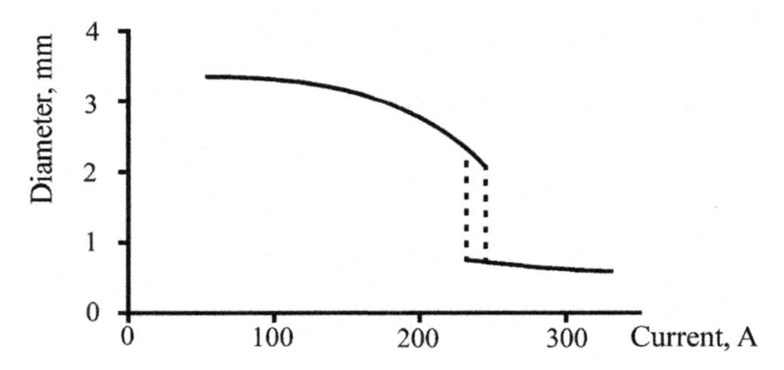

Figure 4.9 Droplet diameter as a function of welding current.

If, on the other hand, we use an argon-rich shielding gas and increase the current to the threshold value for small droplets, we produce a spray arc. In this process, the magnetic forces act downwards (Figure 4.6) and assist droplet nip-off. Plasma currents assist droplet transfer and accelerate them towards the workpiece. In medium range currents, the droplets are about the same size as the wire diameter (Figure 4.8 c). When welding steel wires above the threshold value, the molten metal is often drawn out along the conical tip of the wire, to produce very small droplets (see Figure 4.8 d).

If the welding current is very high, and there is a long wire stick-out, the wire becomes so hot that it has almost melted before it is heated by the arc. This produces an unstable wire end, with the molten metal bending off to the side and starting to rotate. This type of rotating arc (Figure 4.8 e), was originally regarded as unsuitable for welding, but is now sometimes used in mechanised welding of massive joints, and is known as TIME or Rapid Melt Welding.

4.4 Magnetic arc blow

The force or 'arc blow' that arises when the magnetic field around the arc is not completely symmetrical, is a well-known problem with arc welding. In critical cases, it can result in a defective weld. Arc blow can produce the following problems:

- The weld pool, and thus the weld bead, can be deflected towards one side, producing a defective weld.
- If the arc is deflected along the joint, the width of the bead and the penetration can be affected.
- The protection provided by molten slag or gas can be affected, resulting in the formation of pores.

The problem becomes worse, and more noticeable, as the welding current increases. This results in a corresponding rapid increase in all the electromagnetic forces in and around the arc. A number of possible causes of arc blow are discussed in more detail below.

The return current connection is asymmetric

Welding close to a return current connection, or with an asymmetrically connected connection, is a common cause arc blow.

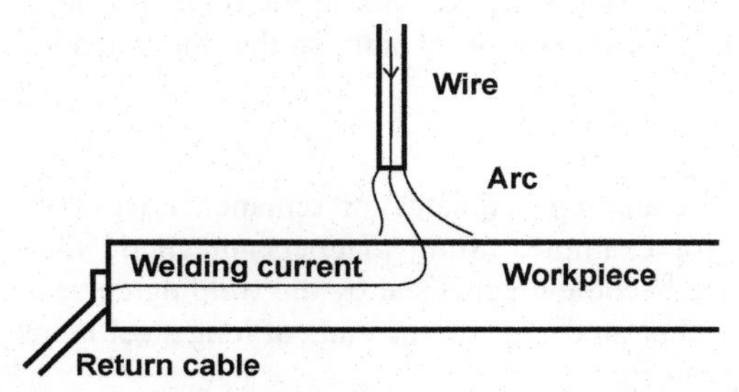

Figure 4.10 Rule of thumb no. 1: The magnetic forces from the welding current attempt to widen the current path.

The workpiece is asymmetric

Magnetic arc blow that can arise when welding close to an edge or where the metal thickness increases (see Figure 4.11).

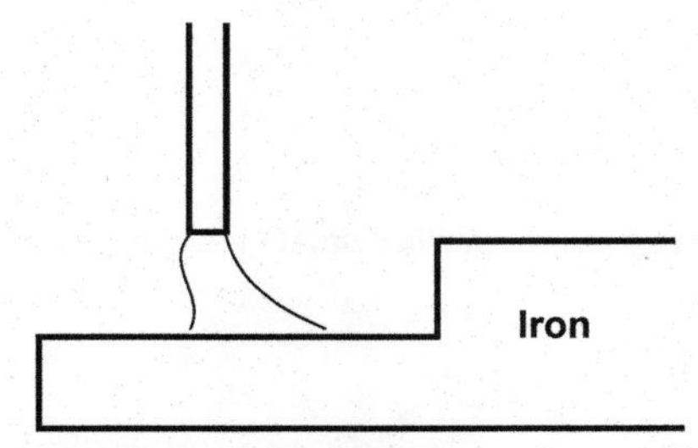

Figure 4.11 Rule of thumb no. 2: If magnetic material (iron) in the workpiece is asymmetrically distributed, the arc will move in the direction where there is the most metal.

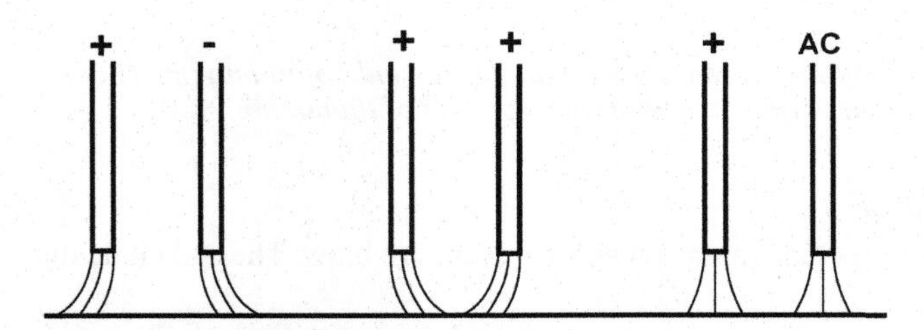

Figure 4.12 Effect from a nearby electrode.

Electrodes close to each other when using multi-wire welding

This problem is common in connection with, for example, submerged arc welding. Each current-carrying conductor is surrounded by its own magnetic field. The magnetic field from one electrode can interfere with the arc from an adjacent electrode (see Figure 4.12).

Induced magnetic fields from the welding current

When welding in steel, the workpiece can provide a path for the magnetic field. An example of this occurs in connection with internal longitudinal welding of a pipe or tube, where the welding current supply cable induces a magnetic flux in the tube. The joint produces a break (also known as an air gap) in the magnetic path, so that the magnetic flux spreads out and affects the arc.

Permanent magnetic fields

These are magnetic fields from magnetic clamping bedplates, or remanence (residual magnetisation) in the workpiece from, for example, lifting magnets, magnetic non-destructive testing or parts of jigs that have become magnetised by the welding current. Even the earth's magnetic field can be concentrated close to the ends of long steel items lying in a north-south direction, affecting the arc.

Alloy steels have a greater tendency to retain residual magnetisation, and cause more problems than soft iron.

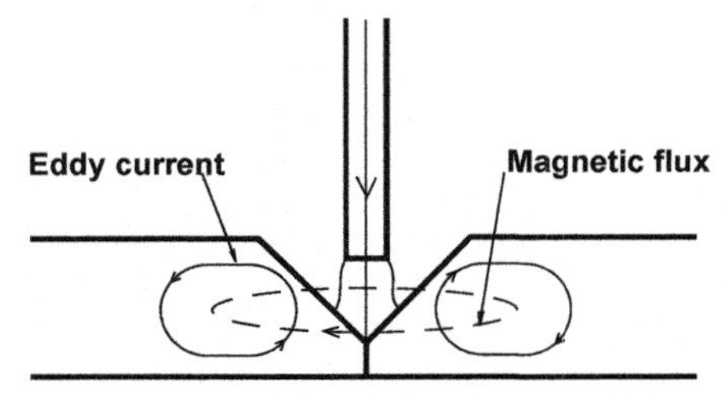

Figure 4.13 Eddy currents in the workpiece limit the magnetic flux at AC welding.

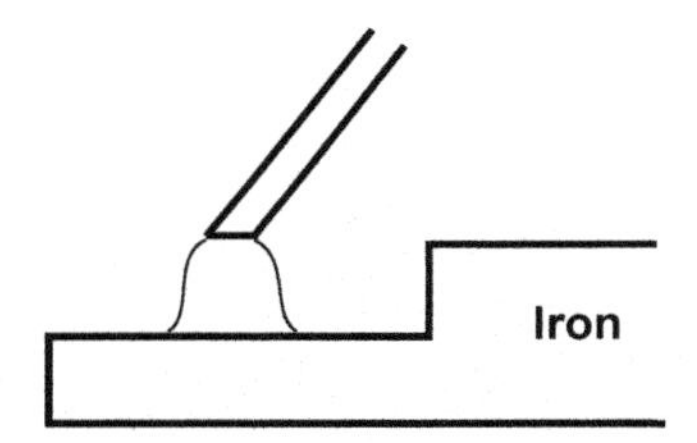

Figure 4.14 Example: Holding the electrode at an angle (see rule of thumb no. 1) can compensate for the arc blow on asymmetric workpieces (rule of thumb no. 2).

Recommended measures

There are a number of measures that can be taken to prevent arc blow. These include the following:

- Do not connect the return current connector close to the position of the weld. Welding towards the return current connection is often preferable. When welding long items, the current can be more evenly distributed by attaching equally long return current cables to each end of the object.
- The use of adequately sized starting and finishing discs can reduce problems at the beginning and the end of a joint.
- AC welding is often better than DC welding: the interference from an external magnetic field is symmetrical, due to the constantly changing direction of the current, and there is less risk of interference resulting from induced fields. This is because the constantly reversing magnetic flux is opposed by eddy currents in the workpiece.

4.5 Shielding gases

The most important reason to use a shielding gas is to protect the molten metal from the harmful effect of the air. Even small amounts of oxygen in the air will oxidise the alloying elements and create slag inclusions. Nitrogen is dissolved in the hot melted material but when it solidifies the solubility decreases and the evaporating gas will form pores. Nitrogen also forms nitrides that may be a cause of brittleness. The shielding gas also influences welding properties and has great importance for weld penetration and weld bead geometry.

Shielding gases work in conjunction with other components of the welding process to protect the molten metal during welding. In MIG-welding with solid electrodes, for example, the deoxidation is partly dependent on the electrode composition and partly on the shielding gas. The electrode contains elements e.g. silicon which react with oxygen and form oxides in the slag that normally appears as small islands on the surface. If flux-cored electrodes are used there are also special slag-forming elements in the flux. The combination of electrode composition and shielding gas improves the quality of the final weld.

Another important role of the shielding gas is to improve other aspects of the process. Some of these factors are

- Ignition of the arc
- Arc stability
- Material deposition
- Wetting between solid material and the weld pool
- Penetration depth and shape
- Spatter formation
- Emission of airborne pollutions

Electrical and heat conduction in the arc plasma depend on the gases used. In the plasma, gases with two or more atoms in the molecule are dissociated into atoms. The atoms are then ionised as the temperature gets higher. The energy needed to form the plasma with atoms and ions is regained at the arc root and melts the material and filler material.

The properties of individual shielding gases

From a chemical point of view the shielding gases influence the formation of oxides in the weld metal and on the surface of the material. Gases also affect other aspects of welding. The properties of the different gases are discussed below.

Argon (Ar)

Argon is one of the most popular shielding gases thanks to its suitable properties. As an inert gas it has no chemical interaction with other materials. It is suitable for all types of welding materials, even more sensitive materials such as aluminium and stainless steel. With MIG welding of mild steel an addition of CO_2 or a small amount of oxygen will improve welding properties, especially for short arc welding. The inclusion of up to

20 % CO_2 improves weld penetration (which will reduce the risk of lack of fusion) while 5–8 % will give reduced spatter.

Helium (He)

Helium like argon is an inert gas. It gives more heat input to the joint which results in a wider and deeper penetration which helps to protects against hot cracking and porosity. Mixed with argon it increases welding speed and is advantageous for penetration in thick-walled aluminium or copper where it compensates for the high heat conduction. Drawbacks with helium are a high cost and low density. A higher gas flow rate must be used to secure the shielding effect. With TIG welding, a high content of helium will reduce ignition properties.

Carbon dioxide (CO_2)

Pure carbon dioxide (CO_2) can be used on its own for short arc welding. It is a cheap gas, suitable for welding galvanised steel and protects against lack of fusion better than argon based gases. Drawbacks are a higher amount of spatter and the fact that the gas cannot be used for spray arc welding.

In the high temperature of the arc, CO_2 dissociates and forms free oxygen atoms. They react with the melted metal. Alloying elements, mainly manganese and silicon, form slag in the weld or on the surface. This decreases the amount of alloying elements in the melted material but is compensated for by a higher alloyed filler material.

Oxygen (O_2)

Oxygen is used as a minor component to stabilise the arc in MIG welding. In this case oxygen is an alternative to CO_2. A higher content of oxygen is avoided as it forms slag. As with CO_2 the forming of slag together with certain alloying elements in steel decreases its mechanical properties.

Hydrogen (H_2)

Small additions of hydrogen can be used to increase heat input and welding speed in the same manner as helium, but at much lower cost. Because of the risk of cracks, hydrogen can only be used for welding of austenitic stainless steel. Owing to the risk of porosity, hydrogen content is limited to a few percent of the total composition of a shielding gas and is normally recommended for welds with one bead only. Hydrogen also actively reduces oxides and is used as a root gas, often in combination with nitrogen.

Nitrogen (N_2)

Nitrogen can be used as an alloying element in duplex stainless steels. A small addition of nitrogen in the shielding gas compensates for losses when welding. Nitrogen or nitrogen mixtures can be used to increase the heat input in the welding of copper, since it has no negative metallurgical effect on copper. Nitrogen is also used as a root gas, often in combination with hydrogen.

4.6 Standardisation of shielding gases

In the European standard EN ISO 14175 *Welding consumables - Gases and gas mixtures for fusion welding and allied processes* the shielding gases for different methods are classified in accordance with their chemical properties. The required gas purities and tolerances of gas mixtures with two or more gases are also specified.

The gases included in EN ISO 14175 are argon, helium, carbon dioxide, oxygen, nitrogen and hydrogen. Their physical and chemical properties are summarized in Table 4.3. Only helium and hydrogen are considerably lighter than air. Carbon dioxide and argon are much heavier than air. At room temperature all are gaseous. In a compressed state normally (200 to 300 bar in cylinders) all of them remain gaseous except carbon dioxide, which is liquid above about 50 bar at 15° C.

TABLE 4.3 Properties of gases used in fusion welding (EN ISO 14175).

Type of gas	Chemical symbol	Specified at 0 °C and 1.013 bar (0.101 MPa)		Boiling point at 1,013 bar	Reaction behaviour during welding
		Density (air = 1,293) kg/m^3	Relative density to air	°C	
Argon	Ar	1.784	1.380	-185.9	Inert
Helium	He	0.178	0.138	-268.9	Inert
Carbon dioxide	CO_2	1.977	1.529	-78.5[1)]	Oxidizing
Oxygen	O_2	1.429	1.105	-183.0	Oxidizing
Nitrogen	N_2	1.251	0.968	-195.8	Low reactive
Hydrogen	H_2	0.090	0.070	-252.8	Reducing

[1)] Sublimation temperature (solid to gas transition temperature).

Argon and helium are inert, which means that they do not react with metals. Oxygen and carbon dioxide are both oxidizing. Carbon dioxide dissociates in the welding arc and forms oxygen atoms which has an oxidizing effect. It is about half as reactive as oxygen. Nitrogen-containing mixtures in the EN-standard are used for root shielding mainly in TIG-welding. Minor additions of nitrogen are also used in MIG-welding for certain stainless steels (see below). Nitrogen is characterized as non-reactive but, at the high temperatures in the arc, it is also dissociated and reacts with the weld pool and the hot material. Hydrogen is a reducing gas at higher temperatures. This means that oxides can be dissolved or prevented from being formed, but its use is limited to certain materials.

Mixtures are often used to combine the effects of different gases. In the EN-standard the classification is based on their chemical reactions.

The different groups are

- I: inert gases and inert gas mixtures
- M: oxidizing mixtures containing inert gases together with oxygen, carbon dioxide or both
- C: highly oxidizing gases and gas mixtures
- R: reducing gas mixtures
- N: low reactive gas or reducing gas mixtures, containing nitrogen
- O: oxygen
- Z: gas mixtures containing components not listed

TABLE 4.4 Classification of shielding gases for arc welding (EN ISO 14175)

Symbol		**Components in percent volume**					
Main group	Sub-group	Oxidizing		Inert		Reducing	Low reactivity
		CO_2	O_2	Ar	He	H_2	N_2
I	1			100			
	2				100		
	3			balance	$0,5 \leq He \leq 95$		
M1	1	$0,5 \leq CO_2 \leq 5$		balance [a]		$0,5 \leq H_2 \leq 5$	
	2	$0,5 \leq CO_2 \leq 5$		balance [a]			
	3		$0,5 \leq O_2 \leq 3$	balance [a]			
	4	$0,5 \leq CO_2 \leq 5$	$0,5 \leq O_2 \leq 3$	balance [a]			
M2	0	$5 < CO_2 \leq 15$		balance [a]			
	1	$15 < CO_2 \leq 25$		balance [a]			
	2		$3 < O_2 \leq 10$	balance [a]			
	3	$0,5 \leq CO_2 \leq 5$	$3 < O_2 \leq 10$	balance [a]			
	4	$5 < CO_2 \leq 15$	$0,5 \leq O_2 \leq 3$	balance [a]			
	5	$5 < CO_2 \leq 15$	$3 < O_2 \leq 10$	balance [a]			
	6	$15 < CO_2 \leq 25$	$0,5 \leq O_2 \leq 3$	balance [a]			
	7	$15 < CO_2 \leq 25$	$3 < O_2 \leq 10$	balance [a]			
M3	1	$25 < CO_2 \leq 50$		balance [a]			
	2		$10 < O_2 \leq 15$	balance [a]			
	3	$25 < CO_2 \leq 50$	$2 < O_2 \leq 10$	balance [a]			
	4	$5 < CO_2 \leq 25$	$10 < O_2 \leq 15$	balance [a]			
	5	$25 < CO_2 \leq 50$	$10 < O_2 \leq 15$	balance [a]			
C	1	100					
	2	balance	$0,5 \leq O_2 \leq 30$				
R	1			balance [a]		$0.5 \leq H_2 \leq 15$	
	2			balance [a]		$15 < H_2 \leq 50$	
N	1						100
	2			balance [a]			$0.5 \leq N_2 \leq 5$
	3			balance [a]			$5 < N_2 \leq 50$
	4			balance [a]		$0.5 \leq H_2 \leq 10$	$0.5 \leq N_2 \leq 5$
	5					$0.5 \leq H_2 \leq 50$	balance
O	1		100				
Z	Gas mixtures containing components not listed, or mixtures outside the composition ranges listed. [b]						

a For the purpose of this classification, argon may be substituted partially or completely by helium.
b Two gas mixtures with the same Z-classification may not be interchangeable.

A gas or gas mixture is classified by the main group and the sub-group (see Table 4.4). An example of classification for an argon rich gas with 6 % CO_2 and 4 % O_2 is:

ISO 14175 – M25

The designation is made up from the classification plus symbols for the gases followed by their respective volume percent. The symbols for the gases are:

Ar: argon
C: carbon dioxide
H: hydrogen
N: nitrogen
O: oxygen
He: helium

An example of designation for an argon rich gas with 6 % CO_2 and 4 % O_2 is:

ISO 14175 – M25 – ArCO – 6/4

Table 4.4 shows the classification of gases used in arc welding.

Tolerances of mixtures

The tolerances of the mixtures are also specified in the standard for component concentrations above 1 %. From 1 % up to 5 % the permissible deviation shall not exceed ± 0.5 % from the specified value. Higher component concentrations may not deviate more than ± 10 % of the component concentration.

Purities and dew point

A prerequisite to get the right mechanical properties of the weld and no porosity is to have high purity gases and a low amount of moisture. Even if the gases in cylinders or in a liquid state are properly formulated, the purity must be assured up to the point of use. This includes:

- Tight and dry hoses and tubes
- No leakage in connections
- No leakage of water or air into the welding guns
- Gas flow without disturbance
- Correct gas flow rate
- Right angle of the welding gun to avoid ejection of air into the shielding area

4.7 Standards for wires and rods

Welding involves a wide variety of base materials. There are consumables for non-alloy and fine grain steels, high-strength steels, creep resistant steels, high-alloy steels, Ni-base steels, aluminium and many more.

Manufacturers' designations of filler materials are based on international classification standards, which in turn assume that the welds are made under (as far as possible) standardised conditions which result in the least possible admixture of weld metal with the parent metal. This produces a weld that does not necessarily represent a real weld, but is nevertheless regarded as the best way of ensuring that different products are tested and compared in a consistent manner.

When classification is based on mechanical properties, the procedures differ between Europe and elsewhere. European standards are based on yield strength and average impact strength at 47 J, while North American and Japanese (“Pacific Rim”) standards are based on tensile strength and average impact strength at 27 J. This has led to the development of ISO standards which contains two solutions, an A-side with the European model (equivalent to EN) and a B-side with a solution developed by the U.S. and Japan. To illustrate the nature of the various standards, this chapter concentrates on two of the European (EN) standards for welding consumables, covering wire electrodes and deposits for gas shielded metal arc welding.

EN ISO 14341 - Wire electrodes and weld deposits for gas shielded metal arc welding of non alloy and fine grain steels

The EN ISO 14341 standard specifies consumable requirements for wire electrodes and weld deposits in the as-welded condition for gas shielded metal arc welding of non-alloyed and fine-grained steels with minimum yield strength of up to 500 N/mm^2. One wire electrode may be tested and classified with different gases. The classification of a weld deposit is based on tests of the all-weld metal. The wire electrode is classified according to its chemical composition and mechanical properties, and the shielding gas. A designation can appear as follows: ISO 14341-A-G 46 3 M G3Si1. The classification is divided into separate components:

ISO 14341	= the standard number
A	= classification by yield strength and 47 J impact energy
G	= wire electrode for gas metal arc welding
46	= strength and elongation (weld metal yield strength 460 MPa)
3	= impact properties (47 J at -30 °C)
M	= the shielding gas ISO 14175 – M2
G3Si1	= chemical composition of the wire electrode.

1. The first part is the standard number.
2. A = classification by yield strength and 47 J impact energy
   B = classification by tensile strength and 27 J impact energy
3. The first part of the classification code is the symbol for the gas metal arc welding process: the symbol G.
4. The second symbol (shown in Table 4.5) indicates the strength and elongation.
5. The third symbol (shown in Table 4.6) indicates the temperature at which an impact energy of 47 J is achieved.
6. Symbol number four can either be M or C, and describes the type of shielding gas as set out in ISO 14175. The symbol M, for mixed gases, is used when the classification has been performed with the shielding gas ISO 14175 – M2, but without helium. The symbol A is used when the classification has been performed with shielding gas M13, Ar from ISO 14175.
7. The fifth symbol indicates the chemical composition of the wire electrode used and is shown in Table 4.7; it also includes an indication of the characteristic alloying elements.

TABLE 4.5 Symbol for strength and elongation of all-weld metal.

Symbol strength (N/mm^2)	Minimum yield elongation (N/mm^2)	Tensile strength (%)	Minimum
35	355	440-570	22
38	380	470-600	20
42	420	500-640	20
46	460	530-680	20
50	500	560-720	18

TABLE 4.6 Symbol for impact properties of all-weld metal

Symbol	Temperature for minimum average impact energy of 47 J (°C)
Z	No requirements
A	+20
0	0
2	- 20
3	- 30
4	- 40
5	- 50
6	- 60

TABLE 4.7 Symbol for chemical composition of wire electrodes.

Symbol	C	Si	Mn	P	S	Ni	Mo	Al	Ti + Zr
G0	*	*	*	*	*	*	*	*	*
G2Si	0.06-0.14	0.50-0.80	0.90-1.30	0.025	0.025	0.15	0.15	0.02	0.15
G3Si1	0.06-0.14	0.70-1.00	1.30-1.60	0.025	0.025	0.15	0.15	0.02	0.15
G3Si2	0.06-0.14	1.00-1.30	1.30-1.60	0.025	0.025	0.15	0.15	0.02	0.15
G4Si1	0.06-0.14	0.80-1.20	1.60-1.90	0.025	0.025	0.15	0.15	0.02	0.15
G2Ti	0.04-0.14	0.40-0.80	0.90-1.40	0.025	0.025	0.15	0.15	0.05-0.20	0.05-0.25
G3Ni1	0.06-0.14	0.50-0.90	1.00-1.60	0.020	0.020	0.80-1.50	0.15	0.02	0.15
G2Ni2	0.06-0.14	0.40-0.80	0.80-1.40	0.020	0.020	2.10-2.70	0.15	0.02	0.15
G2Mo	0.08-0.12	0.30-0.70	0.90-1.30	0.020	0.020	0.15	0.40-0.60	0.02	0.15
G4Mo	0.06-0.14	0.50-0.80	1.70-2.10	0.025	0.025	0.15	0.40-0.60	0.02	0.15
G2Al	0.08-0.14	0.30-0.50	0.90-1.30	0.025	0.025	0.15	0.15	0.35-0.75	0.15

* Any agreed analysis not specified in this standard.

EN ISO 14341 - Wires and rods for arc welding of stainless and heat resisting steels

The EN ISO 14343 standard specifies consumable requirements for wire electrodes, strip electrodes, wires and rods for arc welding of stainless and heat resisting steels. A designation can appear as follows: ISO 14343-A - G 19 12 3 L Si. The classification is divided into two components:

ISO 14343	= standard number
A	= classification to nominal composition
G	= wire electrode for gas shielded metal arc welding.
19 12 3 L Si	= chemical composition of the wire or rod.

1. The first part is the standard number.
2. A = classification to nominal composition, B = classification according to alloy type
3. Symbol for the welding process: G = MIG/MAG, T = TIG, P = plasma, S = SAW, L = laser welding.

   The fourth symbol indicates the chemical composition of the wire or rod.

4.8 References and further reading

K. Weman and G. Lindén, *MIG welding guide*, Woodhead Publishing Limited, 2006.

J.F. Lancaster, *The physics of welding,* 2nd Edition, IIW, Pergamon Press, 1986.

5 Power sources for arc welding

5.1 Introduction

The main purpose of the power source is to supply the system with suitable electric power. Power source performance is of vital importance for the welding process, including features such as the ignition of the arc, the stability of the transfer of the melted electrode material and the amount of spatter generated. For this purpose it is important that the static and dynamic characteristics of the power source are optimised for the particular welding process.

5.2 Electrical characteristics and their control in welding

Static characteristics

The static characteristics of a power unit can be plotted by loading the power unit with an adjustable resistive load. We speak of *drooping characteristics, constant-current characteristics* and *straight characteristics (constant-voltage characteristics)*. These are illustrated in Fig. 5.1.

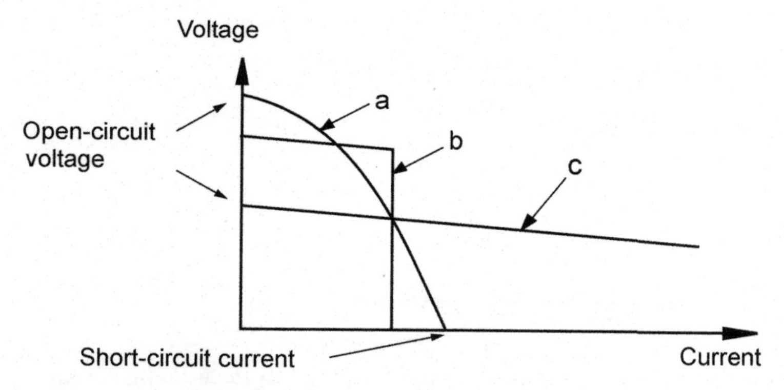

Figure 5.1 Examples of a) a drooping characteristic, b) a constant current characteristic and c) a constant voltage or slightly drooping characteristic.

A *constant-current characteristic* is used when the arc length is controlled by the welder, e.g. in TIG welding. If the arc length is unintentionally changed, the arc voltage changes to maintain a constant current.

A *drooping characteristic* is used for MMA welding, where it is an advantage if the short-circuit current is somewhat higher than the normal load current in order to prevent the electrode from 'freezing' to the workpiece when attempting to strike the arc. A drooping characteristic, as compared with a straight characteristic, also permits a higher no-load voltage which is needed when welding with AC in order to prevent the arc from extinguishing too easily.

If the voltage remains almost constant when it is loaded, it is known as a *constant voltage* or flat characteristic. Typically a voltage drop of 2–5 V/100 A is normal. A flat characteristic maintains good control of the arc length when welding with methods involving a continuously fed filler wire, such as MIG or submerged arc welding. In this case, the current is determined by the speed of the filler wire (i.e. the quantity of filler material being fed into the weld).

Self-regulation of the arc

The point of intersection between the arc characteristic and the power unit load characteristic is referred to as the *working point*. The working point at any particular time represents the welding current and voltage at that time. If the arc length is to be stable in a constant wire-feed system, the power source characteristic must not slope too much.

If, for example, something happens so that the length of the arc is reduced, the voltage drops and the current increases. It can be seen from Figure 5.2 that the current increases from working point 1 to working point 2 if the slope of the characteristic is slight, but only to working point 3 if the characteristic has a steep slope. The increase in current raises the rate of melting of the wire, and the arc length is restored. This is known as the self-regulation characteristic of the arc length. MIG/MAG power units have a straight characteristic in order to provide good self-regulation performance.

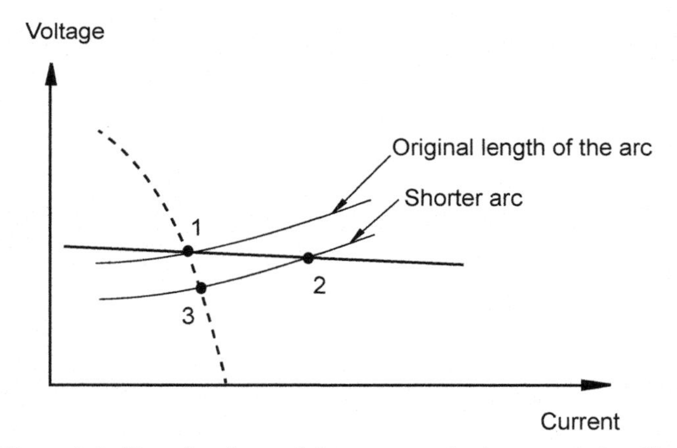

Figure 5.2 How the slope of the power unit characteristic affects the welding current if the arc length is altered.

Setting the current and voltage

When *welding with coated electrodes*, or when performing *TIG welding*, it is the **current** that is set on the power unit, with the arc voltage then depending on the arc length that is used.

When *welding with a continuously supplied filler wire,* e.g. MIG/MAG welding, it is the **voltage** that is set on the power unit. The voltage then determines the length of the arc. This is a result of the arc's self-regulation characteristic. If the welder raises the welding torch, the arc length does not alter: instead, it is the wire stickout that alters.

The current cannot be set directly: instead, it depends on the wire feed speed (and wire diameter) used (see Figure 5.3).

The current, in other words, sets itself so that it is at just the value needed to melt the filler wire at the same rate as the wire is fed out. Changing the voltage, for example, does not greatly affect the current.

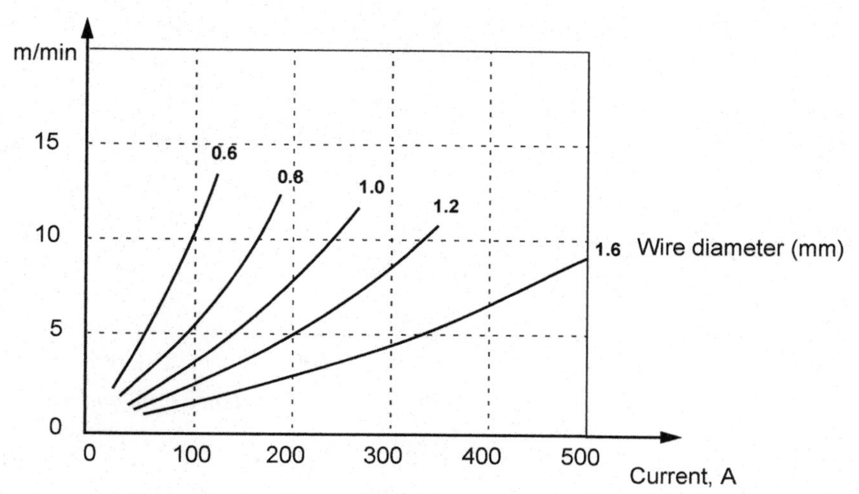

Figure 5.3 The relationship between current and wire feed speed for MIG/MAG welding with normal stickout. (Filler wire: ESAB Autrod 12.51).

When performing submerged arc welding, and some other welding processes, with thicker wires, it can sometimes be preferable to use power units with drooping characteristics. The current then depends on the current setting in the power unit: as a result, the setting procedure is the reverse of what is normally the case. As self-regulation does not work very well with a drooping characteristic, an arc voltage regulator is used to control the wire feed speed. As a result, the voltage and the arc length are kept constant.

Dynamic characteristic

With relatively slow changes in the arc, one can assume that the working point follows the power unit static characteristic. However, in the case of more rapid changes, the dynamic characteristics of the power unit (mainly its inductance) increasingly determine how quickly the current can change to suit. This is important, particularly when welding with short-circuiting drop transfer.

Power units for short arc welding usually incorporate an inductor in their output. The action of the inductor can be likened to the effect of a flywheel in a mechanical system; if the voltage changes instantaneously, as when a droplet of molten metal short-circuits the arc, the current will rise more slowly. It is important that there should not be a current surge during the short circuit, as this would result in high electromagnetic forces that would cause spatter and oscillations on the surface of the weld pool.

The aim is to achieve a high, steady short-circuiting frequency, with finely distributed droplets. The arc should strike quickly and cleanly. It is essential that the power unit has the correct dynamic performance.

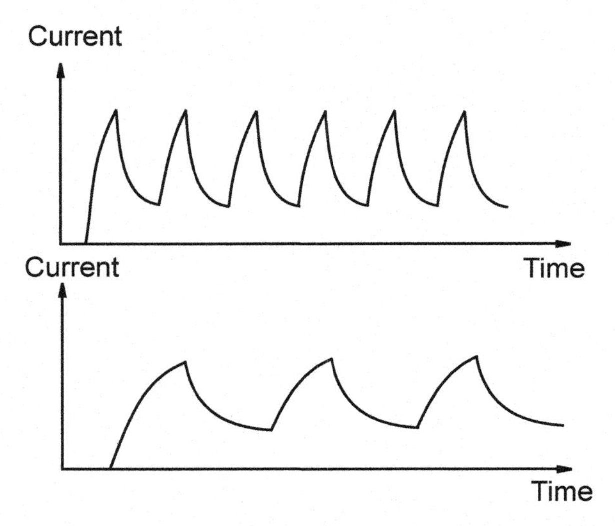

Figure 5.4 Welding current in short arc welding with low inductance (top) and with high inductance.

Welding with alternating current (AC)

AC is a popular choice for welding due to the fact that it uses a simple and inexpensive power unit. Introducing alternating current does, however, lead to complications because, unless special steps are taken, the arc will extinguish on each zero crossing. The need to re-ignite the arc also restricts the choice of coated electrodes and requires a sufficiently high open-circuit voltage of at least 50 V or more. However, electrical safety requirements currently restrict the open-circuit voltage to 80 V. (Special regulations apply in confined or damp areas: see Safety requirements on page 62).

The advantages of alternating current are reduced risk of a magnetic arc blow effect and good oxide-breaking performance when TIG-welding aluminium. AC welding can be a good alternative with certain coated electrodes, as it provides a higher melting rate and reduced smoke generation.

Special power units for AC welding, with a square wave pattern, have been developed. They are electronically controlled and can have rapid zero crossing transitions. This means that they can be used for processes that would otherwise require a DC power source, e.g. TIG or MIG welding. An additional function on these power units is that it is possible to control the relative proportions of the power supply during the positive and negative parts of the cycle, known as balance control.

5.3 Different types of welding power units

The welding power unit converts the high voltage of the mains supply to a non-hazardous level, i.e. it provides a means of controlling the current or voltage and produces the necessary static and dynamic characteristics as required by the welding process. Figure 5.5 shows the historical development of welding power units.

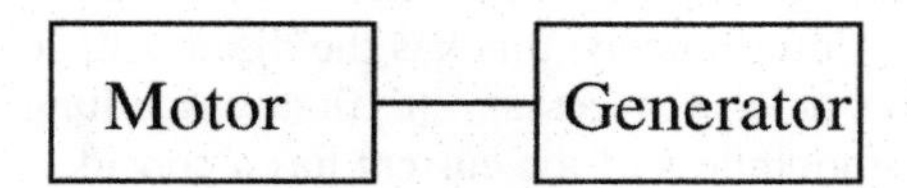

Figure 5.5 The principle of a welding generator.

Motor-generator sets

Motor-generator sets were popular for many years, and are still sometimes used, although no longer manufactured. High cost and poor efficiency made it difficult for them to compete with modern welding power units. However, their welding characteristics can be excellent. They consist of a (3-phase) motor, directly coupled to a DC generator. As the motor speed depends mainly on the mains frequency, these units are relatively insensitive to variations in the supply voltage. They can be remotely controlled by varying the excitation current. Welding generator power units driven by petrol or diesel engines are still made, and fill a need: they are used at sites without a supply of mains electricity.

The welding transformer

Welding transformers provide alternating current, and are the cheapest and perhaps the simplest type of power unit. The standard symbol for a transformer is shown in Figure 5.6. They are used primarily for welding with coated electrodes, although they can also be used with other welding methods when the use of alternating current is required. As opposed to other transformers, welding transformers generally have a drooping characteristic. A common way of achieving this is to separate the primary and secondary windings so that there is a certain leakage of magnetic flux. Adjusting the required welding current is then carried out by moving an additional section of core between the windings in or out by means of a hand wheel. More advanced power units, for use with TIG, submerged arc and occasionally MIG welding, can be controlled by thyristors or transistors using square-wave switching technology. In such cases, it is common that they are able to switch between AC and DC, producing what is known as AC/DC-units.

Figure 5.6 A welding transformer.

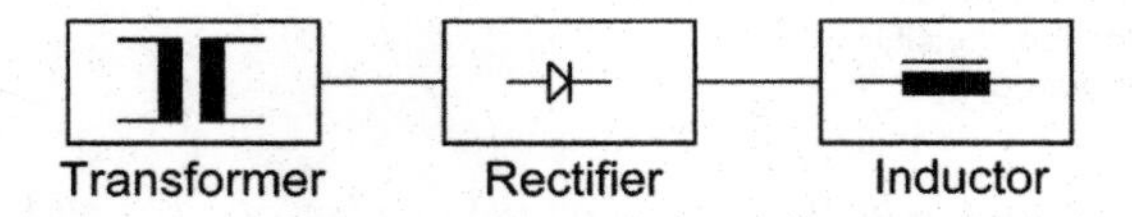

Figure 5.7 A welding rectifier.

Welding rectifier

A traditional welding rectifier power source produces DC, and usually consists of a large 3-phase transformer with some form of rectifier on the secondary side. Power sources having a constant voltage characteristic, e.g. for use with MIG/MAG welding, generally arrange for voltage setting by means of a tap-changer on the transformer. An alternative is to use a thyristor-controlled rectifier bridge. Unfortunately, this has the disadvantage of chopping the output voltage, which makes it also necessary to fit a smoothing inductor (see Figure 5.7). This is because the smoothness of the current has a considerable effect on the welding characteristics.

Thyristor control also provides a means of stepless remote control and insensitivity to variations in the mains supply voltage. Overall efficiency is 70–80 %. The response speed of thyristors is limited by the mains frequency, but is nevertheless sufficiently fast to allow the static characteristics of the power unit to be controlled. This means that the characteristic can be given varying slopes, from straight to drooping, so that the unit can be used with several different welding methods.

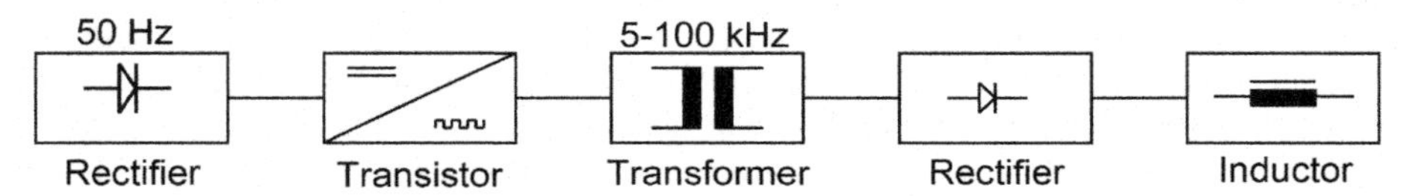

Figure 5.8 A welding inverter.

Welding inverters

Inverter units appeared on the market during the second half of the 1970s. In a primary-switched inverter unit (see Figure 5.8), the 50 Hz mains supply is first rectified and then, using power semiconductors, is turned back into AC at a higher frequency, usually in the range 5–100 kHz. This reduces the weight of the transformer and inductor to a fraction of what is needed for a 50 Hz unit, making the power unit small and portable (see Figure 5.9). Low losses result in high efficiency, to the order of 80–90 %. The high working frequency also allows the unit to be controlled at a speed that is comparable with the rapid processes occurring in connection with droplet transfer in the arc. Such units can therefore have excellent performance. In comparison with traditional power sources, inverter units offer the following advantages:

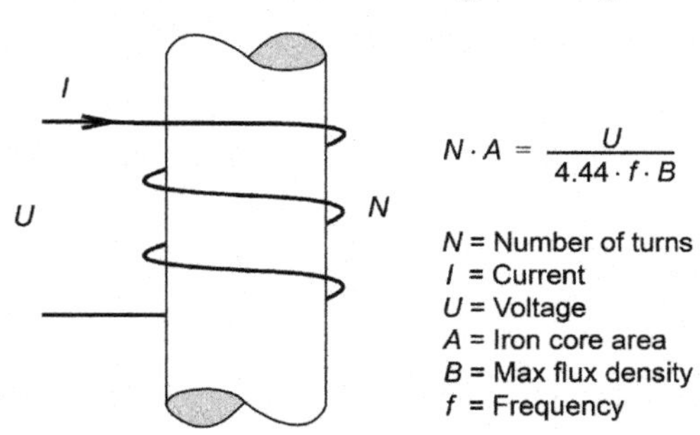

$$N \cdot A = \frac{U}{4.44 \cdot f \cdot B}$$

N = Number of turns
I = Current
U = Voltage
A = Iron core area
B = Max flux density
f = Frequency

Figure 5.9 The size of the transformer and inductor depend on the number of turns and the cross-sectional core area, both of which can be reduced as the frequency is increased.

- Low weight and small size
- Good welding performance
- Several welding methods can be used with the same power source

- High efficiency

A primary-switched inverter power source therefore combines low weight with good control arrangements. Its drawbacks are that it is more complicated and difficult to make adjustable for different mains supply voltages.

5.4 Controlling power sources

Modern electronics and computer technology have had a considerable effect on the development of arc welding equipment. This applies not only to the power circuits, but also to the control electronics in the power unit and in other parts of the electrical equipment used for welding. The following discussion provides a review of the new developments in this area.

Welding characteristics

The welding characteristics of the power source dictate how well the power source performs when welding, e.g. that starting is immediate and without problems, that the arc is stable with a smooth transfer of droplets and that any spatter formation is limited and finely distributed. As a rapidly controllable power source does not essentially have any characteristics of its own, they have to be produced by the electronic or computer control.

Different types of welding methods require different static characteristics. Electronic control increases the flexibility of the power source. It is relatively simple to incorporate control features enabling the power source to be used with several different welding methods. In addition to MIG/MAG, a power source may perhaps also be suitable for use with coated electrodes and TIG welding, without necessarily involving any significant extra cost.

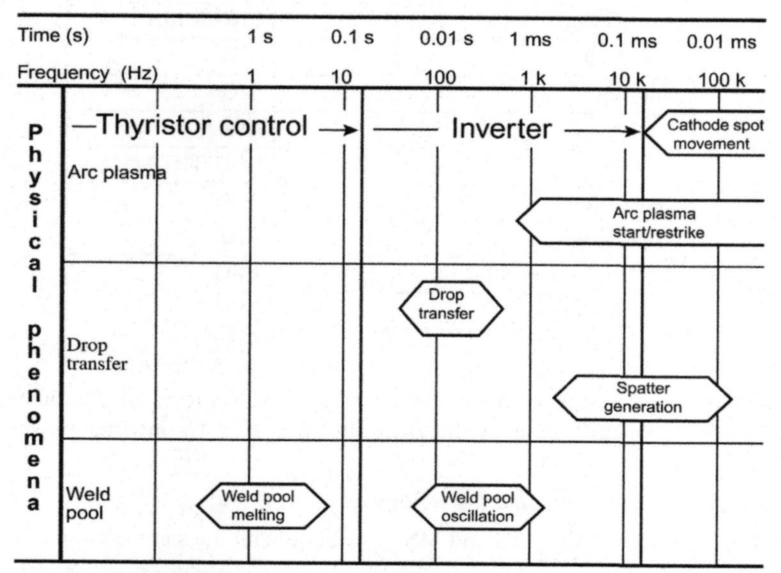

Figure 5.10 The speed of response of the power source is decisive in determining and controlling various processes in the arc.

Control technologies

However, it is not sufficient to simply modify the static characteristic in order to suit a power source to different welding methods. Appropriate dynamic characteristics are needed in order to achieve smooth, stable welding without spatter, particularly when using filler wire processes where the arc is short-circuited by molten droplets (see Figure 5.10).

Good characteristics are particularly important when using short-arc welding, especially when considering the stream of molten droplets to be transferred to the weld pool. Detachment of each droplet is critical, bearing in mind possible spatter formation and the forces that can cause surging of the molten metal in the weld pool. Correct control can maintain a high, consistent short-circuit frequency, resulting in stable transfer of fine droplets with a minimum of large spatter droplets. These characteristics are particularly important when using CO_2 as the shielding gas.

TABLE 5.1 Examples on functions available on advanced power sources.

	Function	Process
Start of welding	Creep start	MIG
	Gas pre-flow	MIG, TIG
	Hot start	MIG, MMA
	HF-start	TIG
	Lift arc	TIG
	Slope up	TIG
Continuous welding	Pulsed MIG welding	MIG
	Arc length control	MIG, TIG
	Step-less inductance setting	MIG
	Synergy lines	MIG, MMA
	Pulsed TIG	TIG
	Slope up	TIG
	Arc force	MMA
	Feedback controlled parameter settings	MIG, TIG, MMA
Finishing a weld	Crater filling	MIG
	Slope down	TIG
	Burn back time setting	MIG
	Shake off pulse	MIG
	Gas post flow	MIG, TIG

The most advanced power sources generally incorporate some form of computer control. It allows maximum utilisation of the flexibility provided by inverter power sources, including:

- Software control of current output and the welding process
- Multi-process possibilities - MIG, TIG and MMA welding with the same equipment
- Synergy line characteristics, providing optimised settings/performance for each situation

- Pulsed arc MIG welding
- Feedback control of welding parameters, guaranteeing improved accuracy and reproduction
- Improved welding start and stop functions
- Man/machine communication with the user through the control panel
- Save and reuse previously used settings by means of a memory

MIG/MAG and other welding processes require several welding parameters to be optimised in order to achieve the best results. A popular way of doing this is the use of single-knob control, known as synergic setting of the welding parameters. This represents combinations of parameters that have originally been established by skilled welders, e.g. combinations of wire feed speed, current, voltage etc., with the results being stored in the memory of the power source. Users start by selecting the required welding method, followed by the type of material, wire diameter and shielding gas. Any subsequent change in the wire feed speed is then compensated by the power source which, at all times, adjusts the other parameters as necessary. Nevertheless, the welder can also override the settings and make manual adjustments from these default characteristics if required. The ability to achieve the intended welding quality is improved by the availability and/or use of various functions, examples of which are shown in Table 5.1.

5.5 Rating data for power sources

The power source rating plate lists the design ratings of the power source, with the most important being the related values of rated current, rated voltage and duty cycle (see Figure 5.11). Other interesting information shown on the rating plate includes efficiency and power factor, open-circuit voltage, insulation class etc.

The International standard IEC/EN 60974-1 specifies requirements for power sources regarding electrical safety. It defines important design principles, rating and testing of the equipment to ensure safe operation.

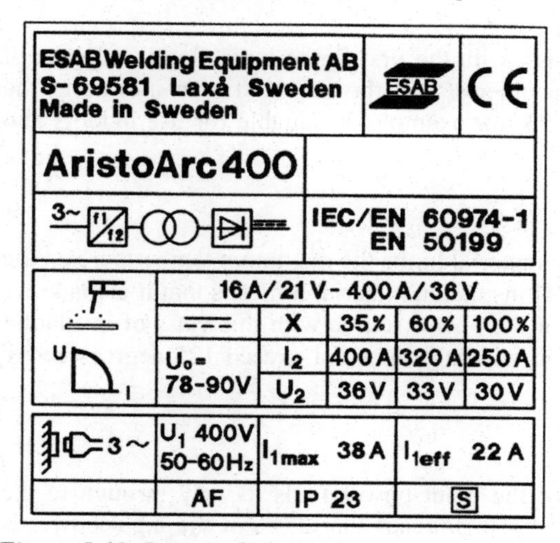

Figure 5.11 Rating plate.

Rated current

The rated current is the current for which the power source is designed. In some cases, a description of a unit may give the impression that it can supply a higher current: always check the technical data or the rating plate to make sure what the actual value of rated current is.

Rated voltage

EN 60974 specifies a standard load line which, for each value of rated current, shows the voltage at which the power source must be tested and with which it must be marked. This means that it is easier to compare the rated data for power sources from different manufacturers. The relationships specified by EN 60974 differ from one welding method to another. Depending on the current, the rated voltages are as follows:

MMA and SAW:	$U = 20 + 0.04 \cdot I$	For currents above 600 A:	U = 44 V
TIG:	$U = 10 + 0.04 \cdot I$	For currents above 600 A:	U = 34 V
MIG/MAG:	$U = 14 + 0.05 \cdot I$	For currents above 600 A:	U = 44 V

Duty cycle

The power source rating is also determined by its duty cycle. This indicates what proportion of a period of ten minutes that the power source can be operated at the specified load. 400 A at 35 % duty factor, for example, means that the power source can supply 400 A for 3.5 minutes in every ten minutes indefinitely without overheating.

Application class [S]

This symbol shows that the power unit is designed for use in areas of elevated electrical risk, i.e. where conditions are cramped (with electrically conducting walls or equipment etc.), or where it is damp.

Enclosure class

The IP code indicates the enclosure class, with the first figure indicating the degree of protection against penetration of solid objects, and the second figure indicating the degree of protection against water. **IP 23**, for example, is suitable for use indoors and outdoors.

Class of insulation

The transformer and inductor insulation material limits the maximum temperature on the windings. If a power source uses class H insulation material it means that it is made for 180°C (20 000 hours). A heating test of the power source with this class of insulation will show that the rise of temperature in windings does not exceed 125 degrees above ambient temperature.

Efficiency and power factor

Efficiency indicates what proportion of the input power finds its way through to the welding process. If the efficiency is 75 %, this means that 25 % of the input power is dissipated in the form of heat losses in the power source. The actual power demand can then be calculated if the efficiency is known:

$$\text{Input power} = \frac{\text{Welding current} \cdot \text{Welding voltage}}{\text{Efficiency}}$$

The active power supplied to the source is measured in kW, and determines the energy cost. The current to be supplied by the mains, and thus passing through the supply fuses, increases if the efficiency is poor. However, in order to be able to work out the supply current, we also need to know the power factor. For a 3-phase supply, we have:

$$I_1 = \frac{P_1}{U_1 \cdot \sqrt{3} \cdot \lambda}$$

where:

I_1 = mains current [A]
P_1 = input power [W]
U_1 = supply voltage [V]
λ = the power factor

The power factor depends partly on the phase displacement between the current and the voltage, and partly on the shape of the current waveform if this departs from a sine wave. Multiplying the current and voltage gives the apparent power, which is measured in kVA and which is of importance when determining the capacity of the electrical supply system.

Typical values of power source efficiency are in the range 0.75–0.85. The power factor can be as high as 0.95, e.g. for a semi-automatic power source with tap-changer control or for certain inverter units, although it is usually considerably lower for MMA power sources. The power factor of large welding transformers with drooping characteristics is often improved by the fitting of phase compensation capacitors, which can improve the power factor from for example 0.40 to 0.70.

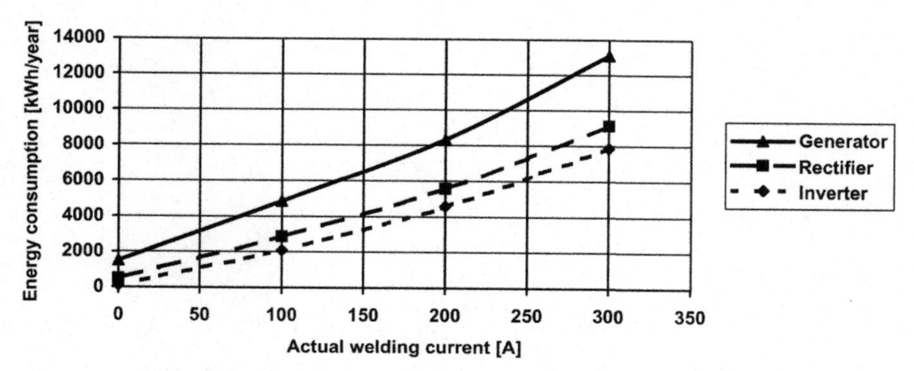

Figure 5.12 Annual energy consumption for different types of manual (MMA) power sources. The differences are due to the different efficiencies and no-load losses of the power sources.

5.6 Safety requirements

It is important from the point of view of electrical safety that the open-circuit voltage of the power unit is not too high. This is particularly important when using AC for welding, where a high open-circuit voltage is often required in order to ensure a stable arc. At the same time, health and safety requirements are particularly strict in connection with the use of AC. EN 60974-1 permits a maximum of 80 V AC, as compared with 113 V DC. Open-circuit AC voltage may not exceed 48 V in wet areas or confined spaces, which are regarded as presenting a higher electrical safety risk. Special devices intended to reduce the open-circuit voltage are available to allow safe welding without affecting welding characteristics.

A welding circuit is not protectively earthed: therefore it is particularly important that the power source is well insulated in order to ensure that the mains voltage cannot reach the secondary circuits.

Transformer winding insulation is exposed to high temperatures, so the material must be of a suitable insulation class to withstand the temperature. A rise of 10°C reduces the life of the material by half. Therefore it is particularly important to keep the interior of the power source clean in order to maintain adequate cooling performance.

Power sources used outdoors should be designed so that moisture and rain cannot degrade the insulation performance.

Despite all these measures, the welder should still take care: the use of gloves, together with undamaged dry clothing and insulated boots, is recommended.

5.7 References and further reading

K. Weman, *A history of welding*, Svetsaren, Vol. 59 No. 1, 2004.

V. Ponomarev, A. Slivinsky, *Welding arc and MIG/MAG welding transfer*, Kiev, 2003.

P. Jiluan, *Arc welding control*, Woodhead Publishing Limited, 2003.

J. Blunt and N. Balchin, *Health and safety in welding and allied processes*, Woodhead Publishing Limited, 2002.

6 TIG welding

6.1 Introduction

Tungsten Inert Gas (TIG) welding (also called Gas Tungsten Arc Welding, or GTAW) involves striking an arc between a non-consumable tungsten electrode and the workpiece. The weld pool and the electrode are protected by an inert gas, usually argon, supplied through a gas cup at the end of the welding torch, in which the electrode is centrally positioned (see Figure 6.1).

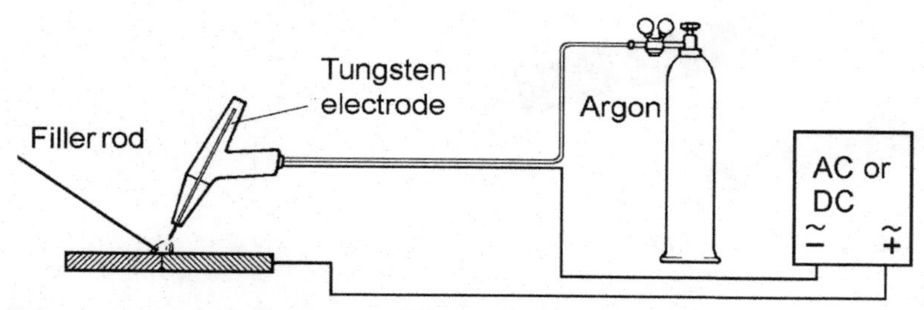

Figure 6.1 Schematic diagram of TIG welding equipment.

TIG welding can also be used for welding with filler material, which can be applied in rod form by hand similar to gas welding. Tools for mechanised TIG welding are used for applications such as joining pipes and welding tubes into the end plates of heat exchangers (see Chapter 15). Such automatic welding tools can incorporate many advanced features, including mechanised supply of filler wire.

The main advantages of the TIG process include the stable arc and excellent control of the welding result. Important applications are welding of stainless steel, light metals such as aluminium and magnesium alloys, and copper. It is suitable for welding all weldable materials, apart from lead and zinc. It can be used with all types of joints and in all welding positions. However, TIG welding is best suited to thin materials, from about 0.5 mm up to about 3 mm thick. In terms of productivity, TIG welding cannot compete with methods such as short arc welding.

6.2 Equipment

The following equipment is required for TIG welding:

- welding torch (including electrode)
- HF (= high-frequency) generator for ignition of the arc
- a power source
- gas tube with regulator
- control equipment

The welding torch

The basic requirements for the welding torch are that it must be easy to handle and well insulated (see Figure 6.2). These requirements apply for manual welding, but are less important for mechanical welding. There are two main types of welding torches: water-cooled and air-cooled. Torches of these two types can carry welding currents of:

- air-cooled: maximum about 200 A
- water-cooled: maximum about 400 A.

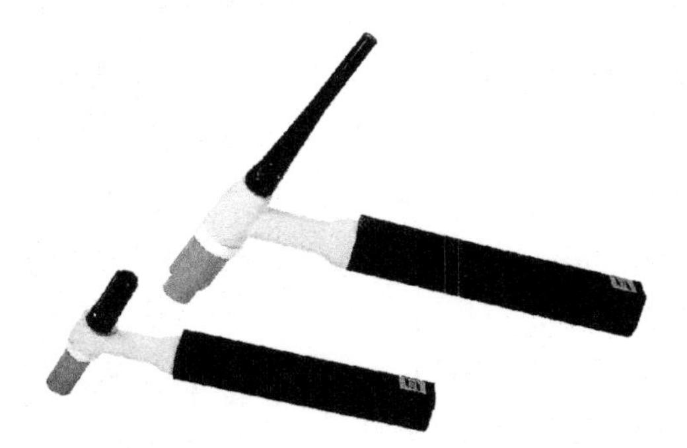

Figure 6.2 Examples of TIG welding torches.

It is important that gas hoses used should be of diffusion-proof material, to prevent moisture, nitrogen or oxygen from the surrounding air from penetrating and contaminating the gas.

The electrode

The electrode material should provide a combination of the following characteristics:

- Low electrical resistance
- High melting point
- Good emission of electrons
- Good thermal conductivity

The material that best meets these requirements is tungsten. It has a melting point of 3370°C and the heat conduction is almost as the same as aluminium.

TABLE 6.1 Examples of ISO 6848 TIG welding electrodes.

Additive	Proportion, %	Colour coding	Type	Current
-	0	green	WP	AC
thorium	2	red	WT20	DC
zirconium	0.8	white	WZ8	AC
lanthanum	1	black	WL10	AC, DC
cerium	2	grey	WC20	AC, DC

Pure tungsten electrodes are used when welding light metals with AC. For other welding applications, the electrodes often incorporate an admixture of 2 % thorium oxide, which improves the stability of the arc and makes it easier to strike. Thorium is radioactive, but is not so dangerous that special precautions are required, apart from taking care when grinding to avoid inhaling the grinding dust. Alternative non-radioactive oxide additives that can be used are zirconium, cerium or lanthanum, as shown in Table 6.1.

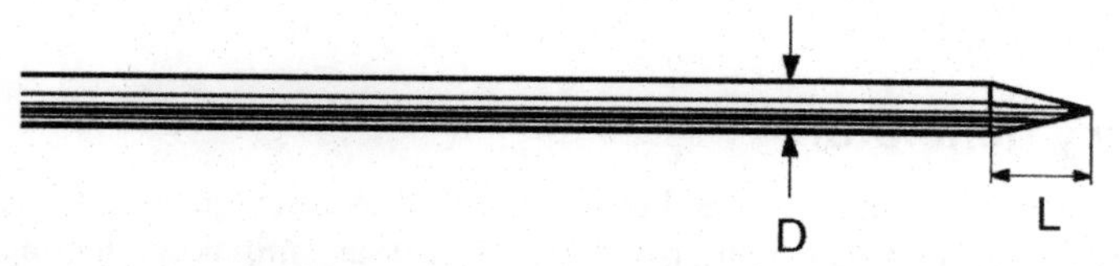

Figure 6.3 Normally the tip of the electrode is ground to a length L = 1.5–2 times the diameter (D).

For DC welding, the tip of the electrode is ground to an approximate 45° angle (Figure 6.3). The use of a special electrode grinding machine guarantees this angle is always the same, as this would otherwise affect the arc and its penetration into the workpiece material. The best arc stability will be obtained if the grinding grooves run the length of the electrode. The tip of the electrode must be ground off.

Electrodes intended for use with AC welding are not ground: instead, the current is increased until it melts the tip of the electrode into a soft, rounded shape.

The electrode diameter is an important variable. The best arc stability is obtained with a high current load, which means that the diameter should be chosen so that the electrode tip is neither too hot nor too cold (see Figure 6.4).

If there is a need for a prolonged electrode stickout, i.e. if the distance between the gas cup and the tip of the electrode is increased, the protection provided by the shielding gas will be less effective. A 'gas lens' is a wire mesh inside the gas cup which reduces eddies in the gas flow, thus extending the length of the laminar flow of the gas without mixing it with air (see Figure 6.5).

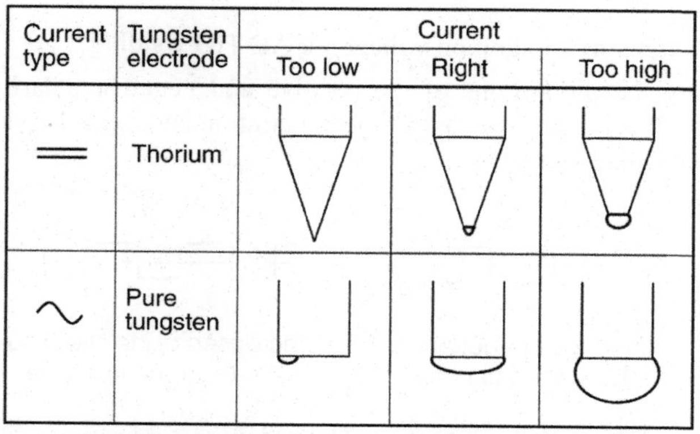

Current type	Tungsten electrode	Current		
		Too low	Right	Too high
=	Thorium			
~	Pure tungsten			

Figure 6.4 TIG electrode tips, showing the effects of too high or too low welding current in relation to the electrode diameter.

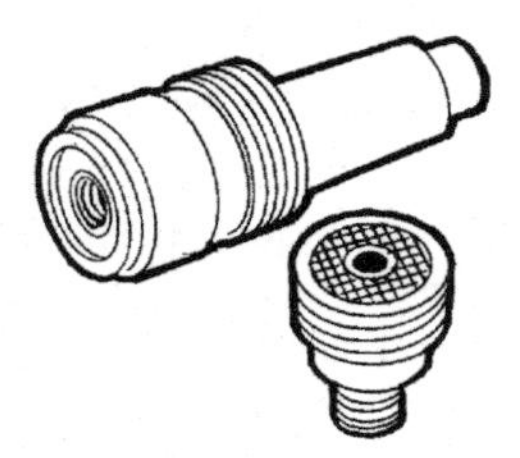

Figure 6.5 Examples of gas lenses.

The high frequency generator

A TIG welding arc is generally ignited with the help of a high-frequency generator, the purpose of which is to produce a spark which provides the necessary initial conducting path through the gas for the low-voltage welding current. The frequency of this initial ignition pulse can be up to several MHz, in combination with a voltage of several kV. However, this produces strong electrical interference, which is the main disadvantage of the method.

The power source

TIG welding is normally carried out using DC, with the negative connected to the electrode, which means that most of the heat is evolved in the workpiece. When welding aluminium, the oxide layer is broken down only if the electrode is connected to the positive pole. However, this then results in excessive temperature of the electrode. As a compromise, aluminium and magnesium are therefore generally welded with AC.

TIG power sources are generally electronically controlled, e.g. in the form of an inverter or a thyristor-controlled rectifier. The open-circuit voltage should be about 80 V, with a constant-current characteristic. When welding with AC (a sine wave), the HF generator is engaged all the time: if not, the arc would extinguish on the zero crossings.

Square wave AC

AC TIG power sources often use a technology involving a square waveform. This means that the zero crossings are very fast, which has the effect of:

- generally not needing a continuous HF ignition voltage for AC TIG welding
- making it possible to vary the proportions of the positive and negative polarity currents, which means that it is possible to control the penetration and oxide breakdown, for example, when welding aluminium.

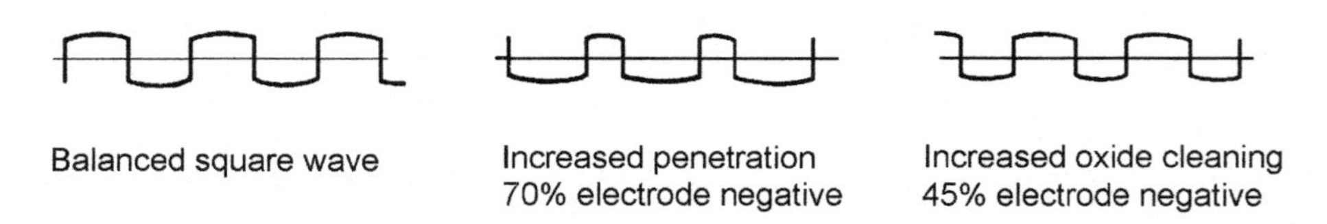

Figure 6.6 Use of a square wave and balance control in TIG welding.

Figure 6.6 shows the current waveform of a square wave supply. The balanced curve normally has 50 % negative polarity on the electrode. Increasing the negative proportion increases the penetration, while reducing it improves oxide break-up. In certain cases, the ability to adjust the polarity setting of the current curve makes it possible to increase welding speed by 50–75 %.

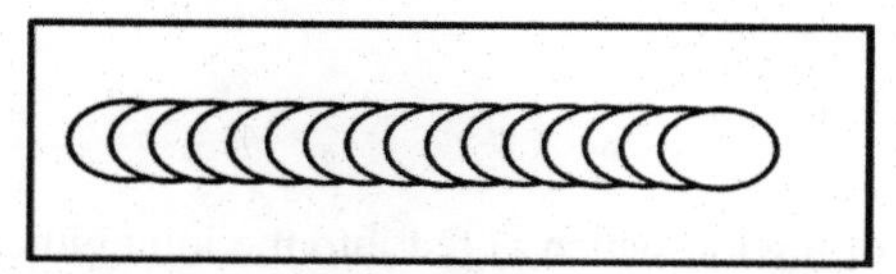

Figure 6.7 The principle for pulsed TIG requires the weld pool to partly solidify between the pulses.

Thermal pulsing

This is used to provide better control of the melt pool and the solidification process. The pulse frequency is set sufficiently low to allow the melt pool to partially solidify between each pulse (Figure 6.7). Supplying the heat in pulses has several benefits:

- Less sensitivity to gap width variations
- Better control of the weld pool in positional welding
- Better control of penetration and the penetration profile
- Reduced sensitivity to uneven heat conduction.

Control equipment

The necessary control equipment depends on to what extent the welding process is mechanised. However, it is usual for the pre-flow and post-flow of the shielding gas, and the HF generator, to be automatically controlled. Crater filling by slope-down of the current, and the ability to pulse the current, are also often employed. Gas pre-flow and post-flow protect the electrode and the weld pool against oxidation (see Figure 6.8).

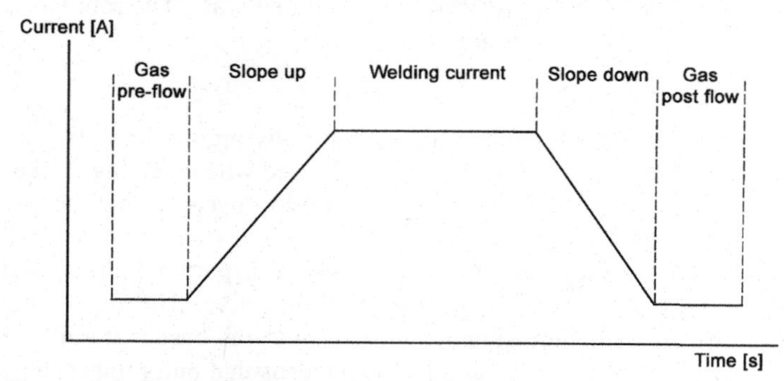

Figure 6.8 Example of a welding sequence.

A-TIG

Coating or spraying the joint surfaces with a thin (10–50 μm) film of an active oxide-containing flux has a surprisingly noticeable effect on penetration: up to three times deeper penetration can be obtained than with ordinary TIG welding, coupled with a narrower weld bead. The benefits are improved productivity as a result of less joint preparation and fewer weld beads. The method, known as A-TIG, has also shown itself to be insensitive to the variations in penetration that can otherwise easily occur due to minor random variations in the chemical analysis of the parent metal.

6.3 Consumables

Filler wire

Fillers for TIG welding are used in the form of a wire, which is fed into the joint either by hand or mechanically. Filler wire for manual welding is supplied in the form of rods about one metre long. It should normally be about the same analysis as that of the parent metal, although sometimes with a slight excess of de-oxidising constituents in order to reduce the risk of pore formation. Thin materials (up to 3–4 mm) can be butt-welded from one side, with the weld metal consisting entirely of molten workpiece material. Higher workpiece thicknesses require some form of joint preparation, with a filler being added in order to fill the joint. The use of fillers is always recommended when welding mild steel in order to reduce the risk of pores.

Shielding gases for different workpiece materials

Steel

Argon is generally used for TIG welding of unalloyed steels, low-alloyed steels and stainless steels. For mechanised welding the shielding is typically argon, with an admixture of helium or hydrogen in order to increase heat input. Hydrogen also helps to reduce oxide formation, and produces a smoother weld, but can be used only on austenitic stainless steel. A small addition of nitrogen may be used when welding duplex stainless steels in order to ensure the correct ferritic/austenitic balance.

When making quality welds with TIG, it is also very common to use a **root gas** in order to protect the root side of the weld against oxidation. This is particularly important in the case of stainless steels or when welding easily-oxidised materials. The root gas is often a mixture of nitrogen/hydrogen, or pure argon.

Aluminium and its alloys

The shielding gas for aluminium and aluminium alloys is usually argon, possibly with the addition of helium. Helium improves heat transfer, and is used when welding thicker sections. The welding current is normally AC or, at low current levels, it may be DC with the electrode connected to the positive.

Under certain conditions, aluminium can be welded with DC if pure helium is used as the shielding gas and the electrode is connected to the negative. The higher arc voltage that results from the use of helium supplies more heat to the base material and thus increases the rate of welding. The high heat input also means that butt joints can be made in thicker sections. The open-circuit voltage of the power source should be sufficiently high to prevent the arc from being extinguished as a result of the higher voltage drop in pure helium.

A drawback of helium is that it is more difficult to strike the arc, particularly with low welding currents. Argon is therefore generally recommended for manual welding, and helium for mechanised welding.

Copper and its alloys
Argon is suitable for welding copper in all positions, and gives excellent results when welding metal thicknesses up to about 6 mm. The high thermal conductivity of the metal generally requires preheating. The best shielding gas for use when welding workpieces more than 6 mm thick is helium, or helium containing 35 % argon.

Titanium
Successful titanium welding requires an extremely high purity of shielding gas, not less than 99.99 %. In addition, extra shielding gas is generally required. Either helium or argon can be used, although argon is generally preferred for metal thicknesses up to about 3 mm due to its higher density and good shielding performance. The use of pure helium is recommended when welding thick sections, due to the resulting higher heat content of the arc.

6.4 Quality issues

Pre and post-treatment
Equipment and materials must be carefully prepared if the highest weld quality is to be achieved. Joints and filler materials must be cleaned from all traces of oil or grease. Brushes must be made of stainless steel. Before welding aluminium or other sensitive materials, it is recommended that they should be degreased with alcohol or acetone, followed by mechanical removal of oxide on joint surfaces by a stainless steel brush, scraping or blasting immediately before welding.

Stainless steel should be handled away from ordinary steel, as any iron particles finding their way on to the metal reduce its rust resistance. Any oxides formed after welding will also reduce corrosion resistance, and should be removed by brushing, grinding, polishing or acid pickling.

Striking the arc
It is not good practice to strike the arc by scraping the electrode on the workpiece: this not only presents risk of tungsten inclusions in the weld, but also damages the electrode by contaminating it with the workpiece material.

Another method of striking the arc is the 'lift-arc' method, which requires the use of a controllable power source. The arc is struck by touching the electrode against the work-piece, but in this case the special power source controls the current to a sufficiently low level to prevent any adverse effects. Lifting the electrode away from the workpiece strikes the arc and raises the current to the pre-set level.

6.5 References and further reading

P. Muncaster, *A practical guide to TIG (GTA) welding*, Woodhead Publishing Limited, 1991.

W. Lucas, *TIG and plasma welding*, Woodhead Publishing Limited, 1990.

7 Plasma welding

7.1 Introduction

The plasma welding method employs an inner plasma gas and outer shielding gas, as shown in Figure 7.1. The plasma gas flows around a retracted centred electrode, which is usually made of tungsten. The shielding gas flows through the outer gas nozzle, serving the same purpose as in TIG welding.

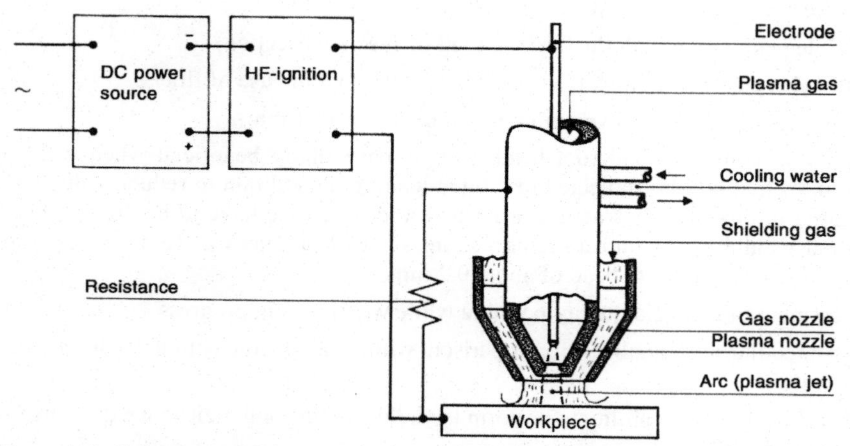

Figure 7.1 Schematic diagram of plasma welding. Resistor R limits the current in the pilot arc which can be ignited also when the torch is apart from the workpiece.

A plasma arc is considerably straighter and more concentrated than, for example, a TIG arc, which means that the method is less sensitive to arc length variations (see Figure 7.2).

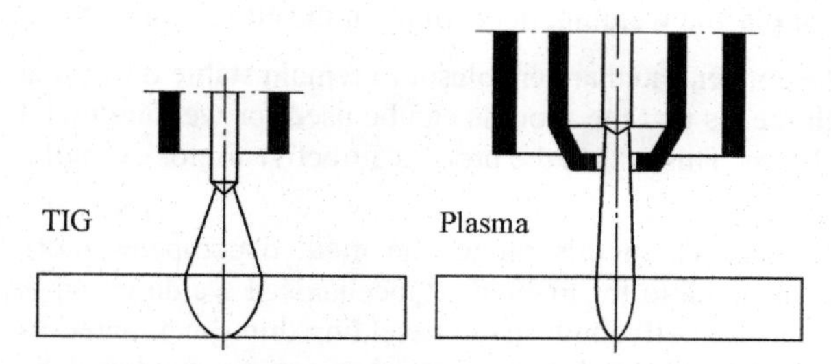

Figure 7.2 The plasma arc is not as conical as the TIG arc, which means that it is much less sensitive to arc length variations.

The plasma welding process can accept variations of 2–3 mm in the arc length without significantly altering the heat input to the workpiece. This is approximately ten times better than the corresponding value for TIG welding. However, because the arc is narrower, more accurate transverse control is important, which means that the method is normally used in mechanised welding.

Advantages and applications

Characteristic features of the method include:

- Reliable arc ignition
- Concentrated highly stable arc at low arc currents with little sensitivity to arc length variations
- High welding speed: up to 400 % higher than that of conventional TIG
- Low heat-affected zone and little distortion
- Reliable penetration with the keyhole method (see next section)
- Butt welds possible in thick materials (8 mm) without the use of fillers.
- Fusion welding possible even in very thin materials (0.1 mm).
- Low weld convexity and root bead. This is particularly beneficial when welding structures that will be subjected to fatigue loading, in addition to reducing the work required in other welds where root bead would otherwise have to be ground away. Plasma welding of 5 mm austenitic stainless steel produces a weld convexity of about 0.3 mm and a root bead of about 0.2 mm.
- Assessment of the weld quality possible while welding is in progress.
- High metallurgical quality in comparison with that of conventional TIG welded materials.
- Flexibility, due to the ability to perform keyhole welding and melt-in welds using the same equipment.

With the exception of magnesium, the method is suitable for welding the same materials as those that can be welded by TIG welding. Automated welding of stainless steel pipes is a major application area.

7.2 Classification of plasma welding methods

There are three different classes of plasma welding, depending on the current range:

- *Micro plasma* (0.1–15 A). The concentrated arc enables it to remain stable down to a current of about 0.1 A, which means that the process can be used for welding metal thicknesses down to about 0.1 mm. This makes the process attractive to, for example, the aerospace industry.
- *Medium plasma welding* (15–100 A). In this range, the method competes more directly with TIG welding. It is suitable for manual or mechanised welding and is used in applications such as the automotive industry for welding thin sheet materials without introducing distortion or unacceptable welded joints, as are produced by MIG welding, or for the welding of pipes in breweries or dairies.

- *Keyhole plasma welding* (>100 A). The third type of plasma welding takes its name from the 'keyhole' that is produced when the joint edges in a butt weld are melted as the plasma jet cuts through them. As the jet is moved forward, the molten metal is pressed backwards, filling up the joint behind the jet (see Figure 7.3).

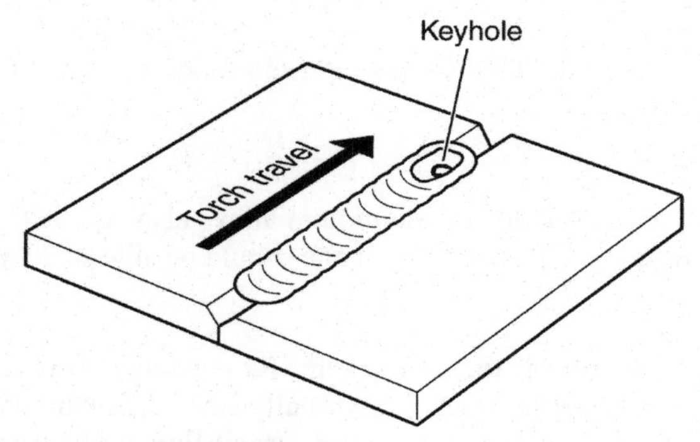

Figure 7.3 Keyhole welding.

The main benefits of plasma welding are to be found in the fact that the keyhole welding method can be used for butt welds from about 3 mm up to 7–8 mm. The keyhole provides a guarantee of full penetration; by comparison the TIG method is only suitable for butt welds up to about 3–4 mm thick. Joints with thicker materials have to be prepared with a V or U joint and then filled with filler material.

Keyhole welding is not suitable for thinner materials below 3 mm. In these circumstances, the process becomes much more like TIG welding. Reducing the plasma gas flow to a low level can make the plasma torch work in the same way as a conventional TIG torch, which can be useful when making tack welds or cladding welds. The main advantage over conventional TIG welding is primarily the excellent arc stability.

There are two types of working systems employed: with transferred and non-transferred arcs, as shown in Figure 7.4.

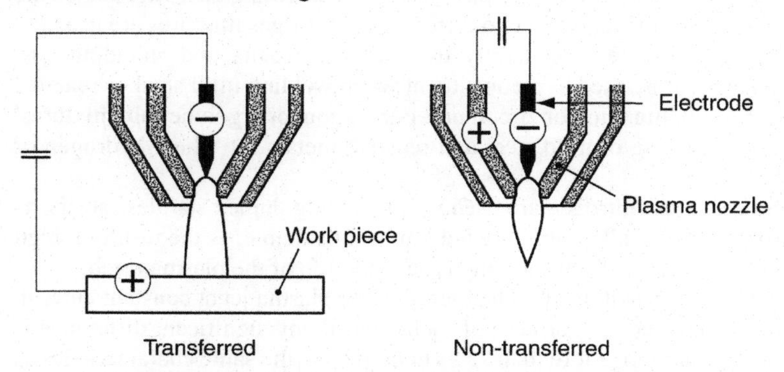

Figure 7.4 Transferred and non-transferred plasma welding arcs.

7.3 Equipment

The following equipment is required for plasma welding:

- Welding torch
- Power source
- Generator
- Control equipment

Welding torch
The same basic requirements apply here as for TIG welding. Plasma welding torches are generally water-cooled.

Power source
Plasma welding employs DC, and for aluminium and aluminium alloys also AC, with a drooping characteristic as for TIG welding. Open circuit voltage should be at least 80 V.

The high frequency generator
In principle, the purpose of the HF generator is the same as in TIG welding. However, when used in plasma welding, the HF generator does not normally strike the main arc. Instead, it strikes a pilot arc as a non-transferred arc, with the current flowing between the electrode and the plasma nozzle. The pilot arc, in other words, can be maintained in air. As the torch approaches the workpiece, the main arc strikes and the pilot arc is extinguished.

Control equipment
The necessary control equipment depends on to what extent the welding process is mechanised. However, it is usual for the pre-flow and post-flow of the shielding gas, the HF generator and the pilot arc, to be automatically controlled. There is often automatic control to ensure that the arc is struck in pure argon, after which the gas supply changes over to the particular gas that is being used.

7.4 Gases for plasma welding

Normally, the same gas is used for both the plasma and the shielding gas. This avoids the risk of variations in the plasma jet when two different gases or gas mixtures are used.

An argon/hydrogen mixture is generally used as the plasma and shielding gas. However, hydrogen cannot be used as a constituent when welding mild steel or reactive metals such as aluminium, titanium, or zirconium. Pure argon, or argon/helium mixtures, are well suited to the welding of mild steel and reactive metals for which hydrogen or nitrogen cannot be used.

Mixtures of argon/helium/nitrogen are used when welding duplex stainless steels, as these contain nitrogen in their alloying. Pure helium is not suitable, as the resulting high heat losses in the plasma gas will substantially reduce the life of the plasma torch.

Argon/helium mixtures result in a higher energy in the plasma jet at constant current. However, the mixture must contain at least 50 % helium if any significant difference is to be noted. Mixtures containing more than 75 % helium have the same characteristics as pure helium.

7.5 References and further reading

W. Lucas, *TIG and plasma welding*, Woodhead Publishing Limited, 1990.

8 MIG/MAG welding

8.1 Introduction

Until the 1970s, manual metal arc (MMA) was the dominant method of welding. Today MIG/MAG is the leading welding process in most industrial countries. Gas metal arc welding (GMAW) is also referred to as MIG (metal inert gas) welding if the shielding gas is inert (e.g. argon) or MAG (metal active gas) welding if the gas has a content of an active gas (such as CO_2).

MIG (and MAG) welding are a particularly flexible methods with a wide range of applications. These include:

- Welding plate thicknesses from 0.5 mm and upwards. The low heat input in MIG welding is particularly useful when welding thin sheet, since it minimises deformation and distortion of the sheet.
- Better productivity in welding thicker metal than many other techniques.
- The ability to weld all commonly encountered structural materials such as mild, low-alloy and stainless steel, aluminium and its alloys, and several other non-ferrous metals (e.g. copper and copper alloys, and nickel and nickel alloys etc.)
- The ability to weld surface coated metals e.g. Zn-coated steel
- Application of the technique in all welding positions.

A limitation of the MIG method compared to MMA is that the welding equipment is more complex and therefore less portable. It has also a more limited application outdoors, as the shielding gas must be protected from draughts.

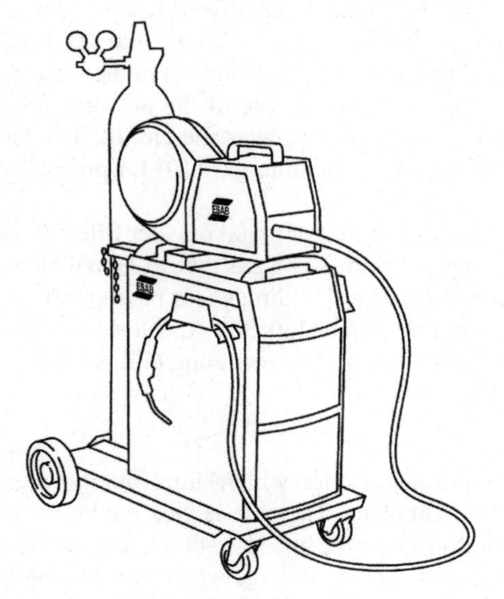

Figure 8.1 Equipment for MIG/MAG welding (ESAB).

8.2 Equipment

Figure 8.1 and 8.2 show the different components of MIG/MAG welding equipment with Figure 8.1 showing the basic unit and how it can be transported to where the welding needs to take place, including the wire feeder and hose which allows the welding gun to be used in any welding position.

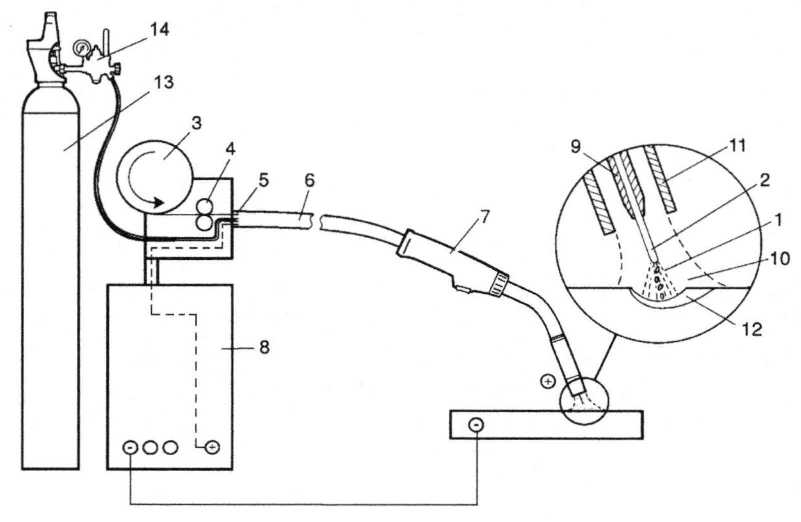

Figure 8.2 The principle of MIG/MAG welding.

Figure 8.2 shows the principle of MIG/MAG welding. The arc (1) is struck between the workpiece and a metal wire electrode (2) that is continually fed forward into the arc. The wire is supplied on a reel (3), and is fed to the welding gun by the drive rollers (4), which push the wire through a flexible conduit (5) in the hose package (6) to the gun (7). Electrical energy for the arc is passed to the wire through the contact tube (9) in the welding gun. This contact tube is normally connected to the positive pole of the power source, and the workpiece to the negative pole. Striking the arc completes the circuit. The gas nozzle (11) that surrounds the contact tube (9) supplies shielding gas (10) for protection of the arc and the weld pool (12).

The wire is normally solid but can also be tubular, a flux or metal powder filled cored wire (see Cored wires on page 82). The gas supply is normally a gas tube (13) to which a regulator and a gas flow indicator (14) is connected. The shielding gas is passed through the hose package to the welding gun. At higher current than 250–300 A, water cooling is often used for the gun and the current supply cable. In this case a cooling unit is needed.

Wire feed unit

The wire reel is placed on a brake hub with adjustable friction which stops rotation when the feeding has stopped in order to keep the wire in place. The wire is then guided to the wire feed rollers (see Figure 8.3). To avoid high friction the hose length must be limited. Most important is to avoid unnecessary curves on the hose package which would rapidly increase the friction.

The wire often has a thin copper coating, which reduces the wear at the contact tip and assists the wire feed, provided of course that the copper layer is properly bonded to the wire. If the wire feed rollers do not press sufficiently firmly on the wire, there is a risk of slipping. On the other hand, if they are set too firmly, they are likely to deform the wire. In either case, there is a risk of particles breaking off the wire and being carried into the wire feed tube. This can easily lead to a vicious circle, with slipping resulting in increased friction, to the point that not even tensioning the feed rollers harder can overcome the friction.

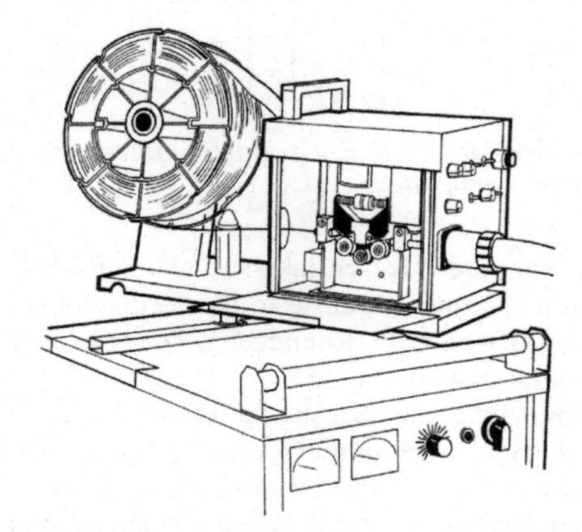

Figure 8.3 Example of a wire feed unit with a wire reel.

The most difficult wires to feed are those that are not copper-plated or which are 'soft', e.g. cored wires or aluminium wires. It is sometimes recommended, when using cored wires, which may not be able to withstand much pressure from the drive rollers, to use tooth-grip grooved rollers for better grip. However, this is harder on the wire, and contributes to increased friction or wear. One way of improving feed performance can be to use several feed rollers: see Figure 8.4.

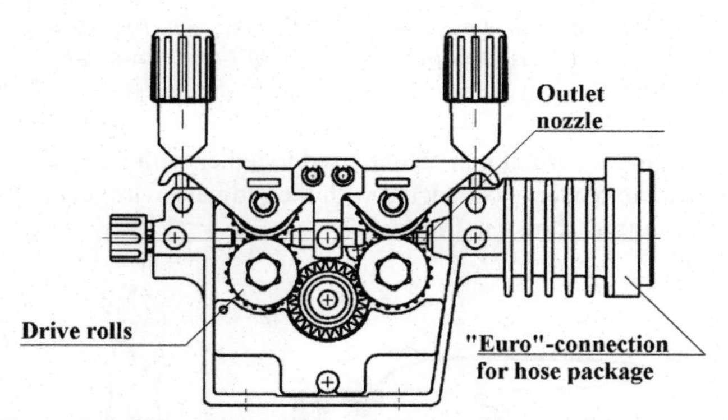

Figure 8.4 Heavy duty 4 drive roll assembly.

Even in normal use it is common for the friction to vary, e.g. when the curvature of the hose is changed or when particles or dirt fill up the wire conduit. The wire speed must not vary too much, otherwise this could result in unwanted variations in the welding data. Superior control of the wire feed speed can be achieved if the motor is equipped with a pulse-generator and feedback system.

The drive rolls have grooves that fit to the wire. It may therefore be necessary to change the rolls when the wire is changed. The number of drive rolls influences the feeding force that can be achieved.

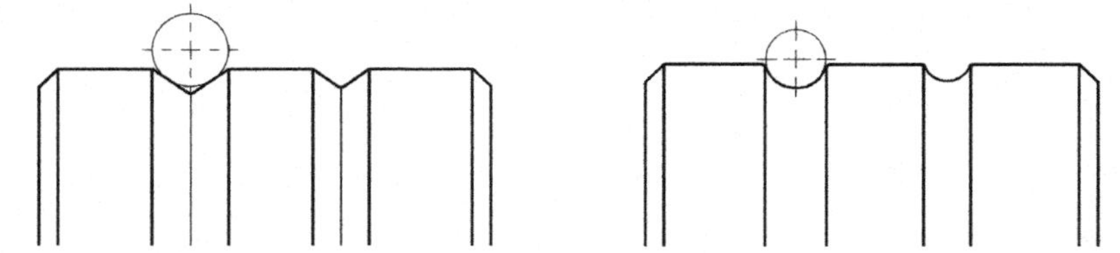

Figure 8.5 Drive rolls with different types of traces are used. For soft wire material as aluminium or tubular wires, U-type traces are recommended.

When welding with aluminium wires special care must be taken to avoid feeding problems. The wire guidance nozzles, shown in Figure 8.4, close to the feeder rollers and the wire conduit need to be of low friction plastic type. Rounded, U-shaped roller traces are recommended for aluminium and tubular wires.

A wire feed system with pushing rolls in the wire feeder is a standard solution that is sufficient in most cases, but for softer wires such as aluminium other solutions are recommended. A push-pull system has a feeding device both in the wire feeder and in the gun as shown in Figure 8.6. It allows also a gun hose length of up to 15 m and is therefore used to increase the operating range for the welder.

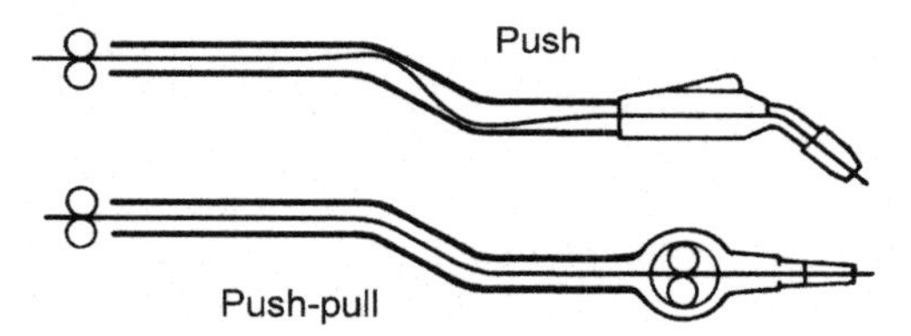

Figure 8.6 The difference between a push and push-pull wire feed system. In a push or pull wire system friction is built up and increased in every curve of the conduit. This is avoided by the use of a push-pull system.

When hose length is important, a normal gun can be used together with an intermediate wire feeder. The intermediate feeder is connected with the ordinary wire feeder by a typically 25 m long hose package as shown in Figure 8.7.

Figure 8.7 An intermediate wire feeder is useful at yards or other places where long reaching is essential (ESAB).

Another solution to feeding problems is a welding gun equipped with its own small wire reel. However, the size and weight of the gun increases and the wire tends to be more expensive. As a result, in Europe the push-pull system is the most popular.

It can be helpful, when welding in higher power ranges, to hang the wire feeder unit above the work station, or to support it and the hose bundle from a counterbalance arm. This reduces the load on the welder and assists wire feed by keeping the hose package in a smooth curve.

The welding gun

The welding gun with hose package is an essential part of the welding equipment (see Figure 8.8). It brings the shielding gas, wire and welding current to the arc. It is difficult to design a robust welding gun for this tough environment and at the same time, make it small and light enough for use in narrow spaces. Welding guns for higher currents are often water cooled. If the current cable, contact tip and gas nozzle is cooled, the size and weight of the gun can be reduced.

The hose package to the welding gun is connected to the gun via a central connector that makes all the connections for the wire, the shielding gas, the welding current and the control wires.

The contact tip of the welding gun is heated from the welding arc, the pool and from the resistive heating of welding current. It can reach temperatures above 300 °C which influences both the hardness and the electrical conductivity of the tip. Special alloys such as CuCrZr are recommended rather than pure copper and show much better hardness and durability at high temperatures. When the contact tip is worn, the point of current transfer is moved back and the conditions for stable welding deteriorate. A clearance between wire and the contact tip bore size is also important. For short arc welding 0.2 mm is employed and for spray arc 0.3–0.5 mm.

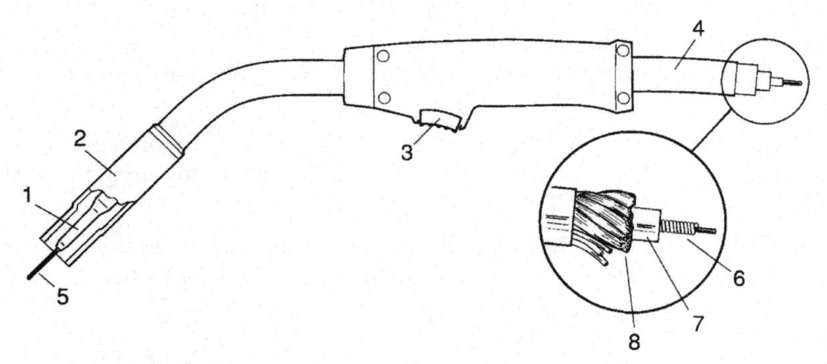

Figure 8.8 Welding gun with hose package. 1. Contact tube. 2. Shielding gas nozzle. 3. Trigger switch. 4. Hose package. 5. Wire. 6. Flexible conduit. 7. Shielding gas hose. 8. Power cable.

Experience shows that careful maintenance is necessary to avoid problems that could occur from constant hot and heavy operation. Aspects of good maintenance include the following:

- At higher welding data the heat from the arc increases. It is important to choose the proper size of gun to avoid temperature overload and to use a water-cooled gun where necessary.
- Keep the gun free from spatter. Spatter will catch easier to a hot surface. Antispatter fluid, which is regularly sprayed on the surfaces exposed to spatter, is recommended to facilitate cleaning of the front parts of the gun.
- Choose a proper wire extension. Too short distance will increase the risk for burn-back of the arc. That will also increase the heat take up from the arc.
- Carefully choose the clearance between wire and the diameter of conduit. Small clearance increases the risk for stoppage and too big will give irregular feeding.
- Follow recommendations from manufacturer cutting the exact length of the wire conduit.
- When feeding problems occur, the reason could be metal particles from the wire increasing friction in the conduit. Increasing the pressure between the feeder rolls is not always the best action. To avoid future problems it is recommended to blow the conduit occasionally to keep it clean.
- Use a dust cover for the wire reel wherever it is exposed to dirt or debris from working environment.
- If using an extra long hose package or soft aluminium wires, a push-pull wire feeder is recommended, see Figure 8.6.
- The sharp edge of the wire tip must always be filed off before feeding the wire into a plastic wire conduit, in order to avoid scratching the surface of the plastic and increasing friction.

Power sources and control systems

DC power sources, with relatively straight characteristics, are used for MIG/MAG welding. There are three main design approaches:

- *Control by a stepping switch.* Tap-changer rectifier units are the most common type used.
- *Thyristor-controlled rectifier units.* These are often larger and are more advanced as they provides means of stepless adjustment and are insensitive to variations in the mains supply voltage.
- *Inverter power sources.* These have the most advanced design. In addition to their good welding characteristics and control facilities, inverters are often used for welding aluminium and stainless steel, which benefit from the use of pulsed MIG welding.

The MIG process has the ability to operate without any electronic control. If the wire is fed with constant speed and the power source is of a constant voltage type, the arc length will maintain stable. This is because of the self-adjusting nature of the arc. If some disturbance happens that changes the arc length, the voltage drop over the arc is changed and the resulting current reaction from the power source influences the rate at which the wire melts such that the arc length will be stabilised again.

The high short circuit current of the power source is an advantage at the start of welding, when the wire hits the workpiece. Normally, a stable arc is maintained within a tenth of a second.

Various control techniques can be used to optimise the process. In short arc welding, the high short circuit current needs to be reduced to avoid spatter. This is traditionally done with an inductor coil, often with different output terminals on the power source to allow the user to optimise process stability. More advanced power sources use electronic or digital control to adjust current amplitude.

Electronic and computer control of the welding equipment can improve weld quality in a number of ways, as described in the chapter on power sources. In addition to functions that improve starting and stopping, there are also several ways of controlling the welding process in real time. The major welding equipment manufacturers have developed their own proprietary solutions, such as Lincoln's STT (Surface Tension Transfer) and Cold Arc from EWM, which assist the short-arc process in various ways by actively controlling the current during the short circuit and normal arc times. Typical benefits of such systems include a cooler welding process, which causes less distortion, improved bridging of gaps, less spatter and making the process very suitable for MIG brazing. Other control systems include:

- ESAB's QSet™ power source which automatically adjusts the arc voltage, while being unaffected by voltage drops in the wire stickout or elsewhere.
- Fronius' CMT (Cold Metal Transfer) system which controls the wire feed very precisely. It can be used for low-energy-input welding, which allows welding of very thin materials and is also ideal for MIG brazing.

On traditional welding equipment the welding voltage will vary as the mains voltage fluctuates. The resistance of the transformer windings increases when they are heated and also causes a voltage drop that may influence welding quality. More advanced equipment is therefore normally equipped with feedback control that maintains the welding parameters set at the start of welding and by this contribute to secure the welding quality (see Figure 8.9).

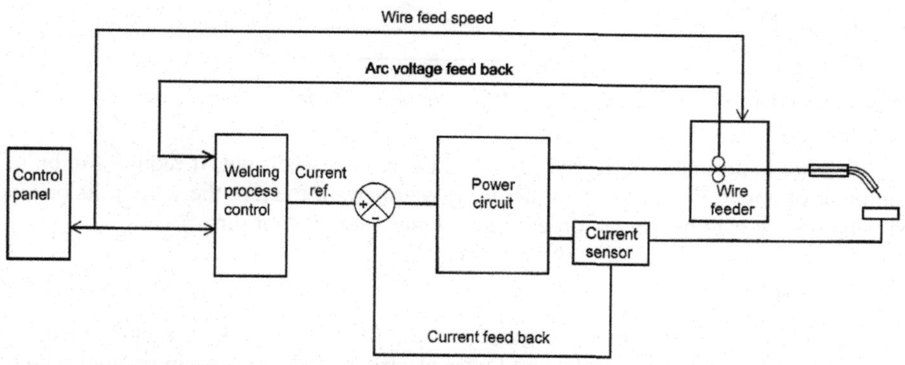

Figure 8.9 The principle for a process control system with feedback control of welding voltage and current. (ESAB).

Cooling units

Water-cooled welding guns are often used in the higher current range (300–500 A). Cooling water is circulated from a cooling unit, which may be separate or be incorporated in the power source. The water cools the copper conductor in the hose and cable bundle, the gas nozzle and the contact tip. Cooling units normally include a water container, a pump and a fan-cooled radiator.

8.3 Consumables

Suppliers' product catalogues provide help in choosing a suitable filler material. As a rule the filler material should have a chemical composition and mechanical strength that is similar to that of the base material.

Solid wires

Solid wires for MIG/MAG welding are available in the 0.6–2.4 mm range for use with many different types of materials. Wires are normally supplied on reels and wound to ensure that the wire does not snag when being withdrawn. Important factors are that it must be clean with a smooth finish and be free from metal flaws to feed easily. Wires coated with a thin layer of copper have the best feeding performance. One important condition is that the copper is well fixed to the wire. If not, it will clog up the wire conduit and prevent smooth feeding.

To get the most effective performance from the arc, it is essential that the current is transferred to the wire close to the opening of the contact tip. To improve the contact force, and to define the contact point, the wire is somewhat curved, i.e. it has a radius of 400–1200 mm (see Figure 8.10). The helix size should not exceed 25 mm, if problems with arc wandering are to be avoided. If maximum reliability is required, as for example at advanced robot welding, copper-free wire may be required.

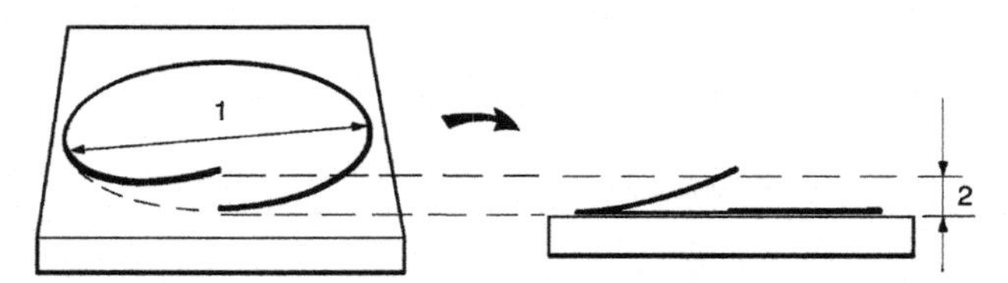

Figure 8.10 Checking cast diameter (1) and helix size (2).

The wire is normally delivered on 10–15 kg spools (steel) but, if required in bulk, a container of about 200 kg can be ordered. It is important to store the wire in its original packing until it is to be used in order to avoid moisture, dust or dirt.

Cored wires

Cored wires consist of a metallic outer sheath, filled with flux or metal powder, as shown in Figure 8.11. The use of cored wire electrodes is increasing in parallel with the introduction of new types of wire. They are used, for example, for:

- welding thicker plates: sheet thicknesses from 4 mm and upwards.
- both butt and fillet welds.
- manual welding in all positions.
- robot welding in the horizontal position.

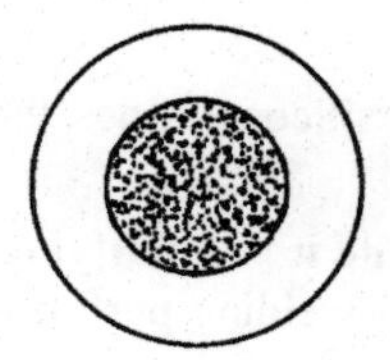 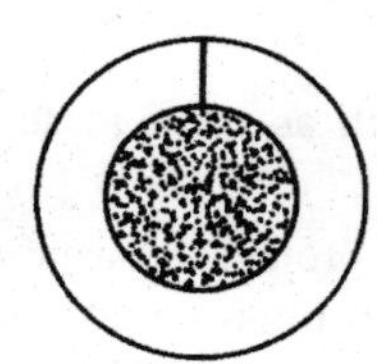 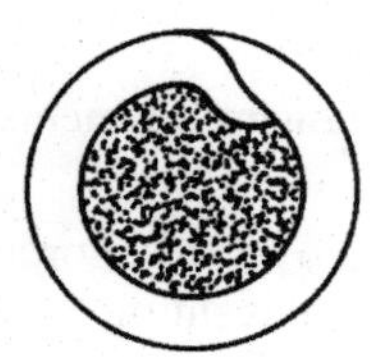

Figure 8.11 A section through different types of cored filler wires. The proportion of filler powder varies depending on differences in the wall thickness.

The benefits of cored filler wires are:

- a higher deposition rate compared to solid wires (see Figure 8.12).
- high productivity in positional welding.
- basic wires have the ability to produce a tough, crack-resistant weld since they are tolerant to contamination in the material.
- better side wall penetration than with solid filler wires.
- less risk for lack of fusion in comparison with solid wires.
- stabilising substances in the powder extend the range of usable welding data.

The drawbacks of cored filler wires are:

- a higher price than solid wires (does not necessarily mean a higher total cost).
- thermal radiation at higher welding currents.
- finishing work required when using slag-forming wires.
- relatively large quantities of fume produced by self-shielded wires.

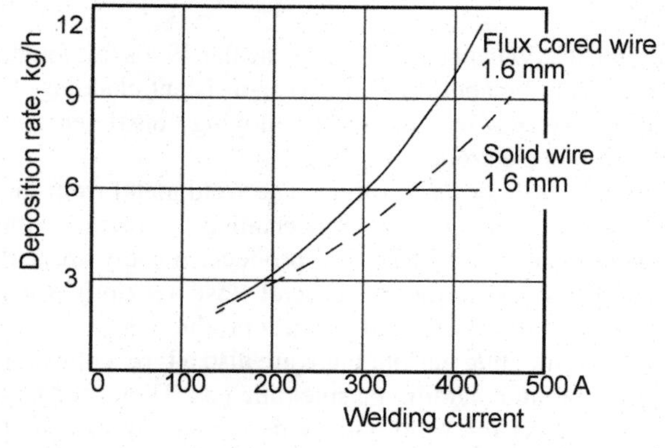

Figure 8.12 Deposition rate when welding carbon steel.

Cored wires are manufactured from 0.8 mm upwards in diameter, with the commonest sizes being 1.2, 1.4 and 1.6 mm. The range of weld metal grades is wide, and is constantly increasing. In particular, the range of cored filler wires for use with stainless steel has increased.

The composition of the powder in cored wires differs for different types of wires and it affects the welding characteristics and the metallurgical analysis of the weld metal.

Wires can be optimised for a range of characteristics by varying the composition of the powder:

- The addition of anti-oxidising elements such as manganese or silicon refines the weld metal.
- Slag-forming elements are added in order to protect the weld while it is solidifying, to control the shape of the weld metal and to improve positional welding performance.
- Arc-stabilising additives produce a stable, spatter-free arc.
- Alloying elements such as nickel, chromium, molybdenum and manganese can be incorporated in the powder in order to modify the mechanical and metallurgical properties of the weld.

Essentially the same equipment is used for cored wire welding as for ordinary MIG/MAG welding. However, the welding gun, the wire feed unit and the power source all need to be more powerful due to the high current and the thicker wire. Welding is usually carried out using DC, with the wire connected to the positive pole. The power source characteristic is generally slightly drooping, which gives a self-regulating arc.

Fume is a problem when welding with high current, not least when using self-shielded flux cored wires. One solution to this problem is to use a welding gun with an integral extraction nozzle.

Metal-cored arc welding (MCAW) and flux-cored arc welding (FCAW) wires

We can distinguish between *metal-cored arc welding* (MCAW) and *flux-cored arc welding* (FCAW) wires. Metal powder-filled cored wires contain a powder that consists mainly of iron and alloying elements. Adding metal powder alone produces a slag-free cored wire that has a higher productivity than a flux-cored wire. The only slag formed is in the form of small islands of silicon oxide. These wires have a high productivity in the horizontal position.

The use of flux-cored arc welding (FCAW) appeared as early as the 1920s, although at that time it was only in connection with the application of wear-resistant cladding. It was easy to produce high-alloy filler materials by mixing the alloying constituents in powder form inside a cored unalloyed steel electrode.

Fluxed-cored wires can have either a rutile or basic filling. The weld metal from the basic wires is of high purity and therefore can show good mechanical properties with very good crack resistance and impact values. The rutile type produces a spray arc and the best welding characteristics. Its slag supports the weld metal (fast freezing) when welding out of position and allows positional welding with highest productivity.

Flux cored wires normally require a shielding gas but there are also a type known as *self-shielded flux cored wires* that do not require additional shielding gas. They are filled with a powder that develops gases to protect the weld pool. This is done by means of appropriate additives which are gasified in the arc. The resulting substantial expansion excludes the surrounding air from the arc and weld pool. When carrying out cored wire welding without a shielding gas, the same power source and wire feed unit are usually used as would be used for welding with shielding gas. However, the welding gun can be simpler, as there is no need for a gas supply.

8.4 MIG/MAG welding process variations

The stability of a DC arc with a consumable electrode (i.e. a filler wire) depends largely on how the molten metal is transferred in the arc. One can distinguish essentially between two different types of arcs, depending on the material transport: the spray arc and the short arc (short-circuiting arc).

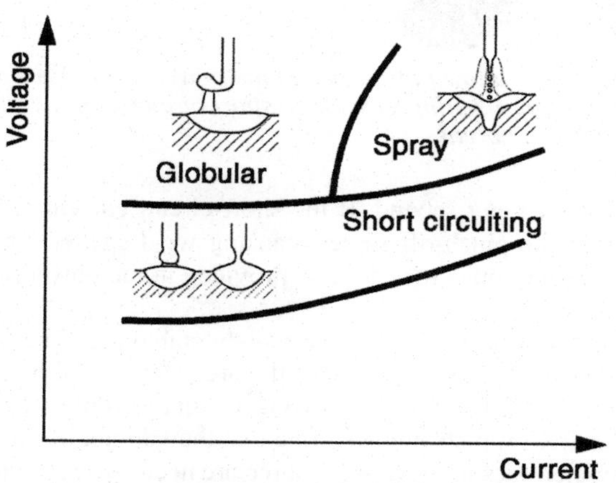

Figure 8.13 Arc types for different current and voltage conditions.

Spray arc welding

Spray arcs are characterised by the transfer of molten material in the form of many small droplets, the diameter of which is less than that of the filler wire. As there are no short circuits, the arc is stable and spatter-free. A prerequisite for successful spray arc welding is that the values of current and voltage should be over certain limits. This, in turn, means that more heat is supplied to the workpiece than with short arc welding, and only materials of 5 mm thick or more are suitable for spray arc welding. The high heat input means that the weld pool is also large, so welding has to be performed in the horizontal position. It should be noted that a pure spray arc cannot be obtained when using CO_2 as the shielding gas. The shielding gas must be pure argon, preferably with a small proportion of CO_2 (not more than 25 %) or O_2. Spray arc welding is particularly suitable for MIG welding of aluminium and stainless steel, for which the shielding gas is mainly argon.

With a thin filler wire, it is possible to perform successful spray arc welding at lower currents than with a thicker filler wire.

The arc voltage should be set just high enough to maintain a short-circuit-free arc. The filler wire is normally connected to the positive pole.

Short arc welding

The heat input from short arc welding is low, which makes the process suitable for welding thinner materials. The drops from the wire dip into the weld pool. The arc is therefore periodically replaced by a short-circuiting bridge of molten metal (see Figure 8.14).

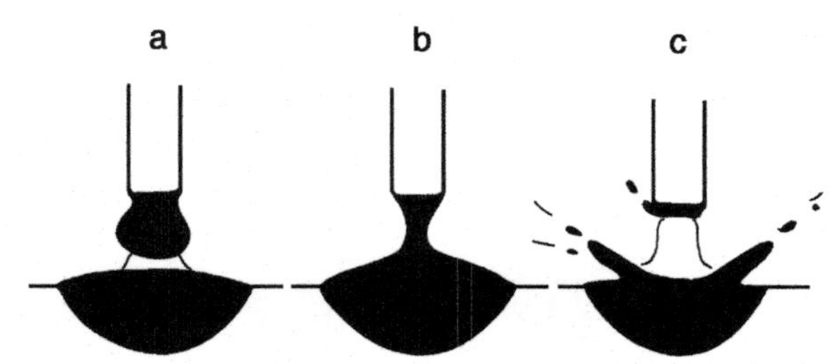

Figure 8.14 Droplet short-circuiting with a low inductance in the power unit. a) Arc period. b) Drop transfer. c) Low inductance setting gives high short circuit current and spatter is developed when the short circuit breaks.

This can be repeated up to 200 times per second. If the short-circuit current is too high, it has a considerable effect on the pinch-off forces, causing weld spatter. Some means of limiting the short-circuit current must therefore be provided in the power unit, e.g. through the use of an inductor coil.

It is not easy, with short-arc welding, to achieve a completely stable arc. The best way to judge if the welding is going well is by the noise of the arc. The objective is to achieve a consistent, high short-circuiting frequency, resulting in small droplets being transferred to the workpiece and spatter droplets being so fine that they do not adhere to the workpiece. Good welding characteristics in the power source are necessary, although wire feed speed, current transfer in the contact tip and the welder's skills are also important.

Ignition of the arc can also be sensitive, and therefore it is important that all parts of the equipment should be in good condition in order to strike the arc successfully.

Globular transfer

At currents lower than needed for spray transfer and with voltage above pure short arc welding, there is a mixed region characterised with droplets larger than the wire diameter and often with an irregular shape. The molten drop grows until it detaches by short-circuiting or by gravity. The globular transfer mode is usually best avoided.

Pulsed MIG welding

Pulsed arc welding is used mainly for welding aluminium and stainless steel, although it can also be used for welding ordinary carbon steel. The method of controlling the transfer of the droplets by current pulses (30–300 Hz) from the power source makes it possible to extend the spray arc range down to low welding data. The process provides a stable and spatter-free arc as an alternative to short arc welding.

Pulses from the power source pinch off the drops from the wire at the same speed as the wire is fed (see Figure 8.15). Therefore it is possible to avoid short circuits and spatter generation.

Modern electronic inverter power sources are able to calculate the pulse shape needed for the actual choice of wire size, material and shielding gas and the pulse frequency needed to keep the arc length constant. The advantages of pulsed MIG welding are:

- The process is fully controlled and spatter-free.

- The ability to extend spray arc welding down to lower welding data is particularly suitable when welding materials such as stainless steel or aluminium. It becomes possible to weld thin materials, or to perform positional welding, with better results than would be obtained with short arc welding.
- Pulsed MIG welding is sometimes used within the normal spray arc range in order to provide better penetration into the material.
- Stable welding performance can also be achieved with a somewhat thicker filler wire that is less expensive and easier to feed. This is particularly useful when welding aluminium, as it is difficult to feed thin aluminium wires due to their softness.
- Research work has shown that the efficient droplet pinch-off reduces overheating of the droplets, resulting in less fume production.

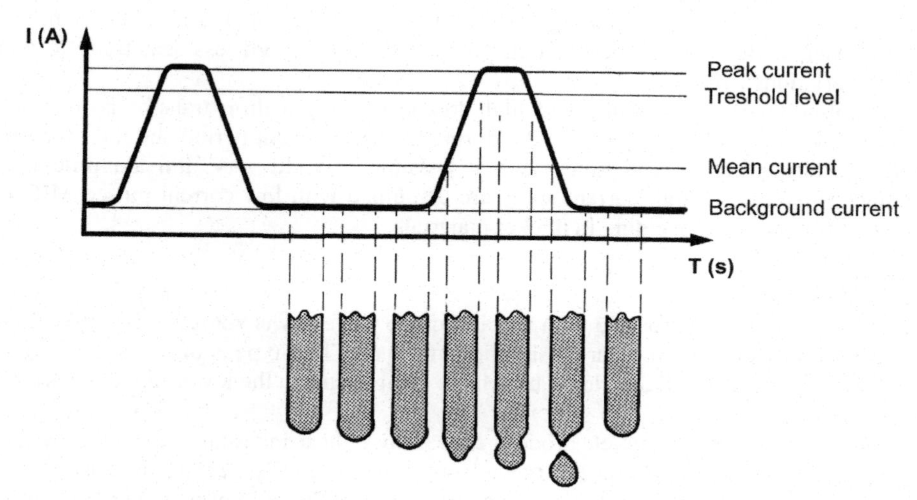

Figure 8.15 Pulsed MIG welding gives a spray arc free from short circuits at a lower welding current that otherwise is possible.

The disadvantages of pulsed MIG welding are:

- Production speed is generally lower than with short arc welding. The greater heat input, relative to that of short arc welding, reduces the maximum usable wire feed speed.
- Pulsed arc welding restricts the choice of shielding gases. As with spray arc welding, the CO_2 concentration of an argon/CO_2 mixture must not be too high.

Thermal pulsing

The low frequency pulsing generally used with TIG welding is called thermal pulsing in order to separate it from the higher frequency pulsed drop transfer process used at MIG welding.

Thermal pulsing is used to achieve improved control over the weld pool. When used for MIG welding the wire feed speed is pulsed with a frequency in the range 1 - 10 Hz. The process involves heating a point on the workpiece until the metal is completely molten. The welding current is then reduced, such that much of the weld pool solidifies, after which the process is repeated as the arc is moved along the joint. The overlapping weld ripple creates a pattern that sometimes is desirable as a styling effect (see Figure 8.16).

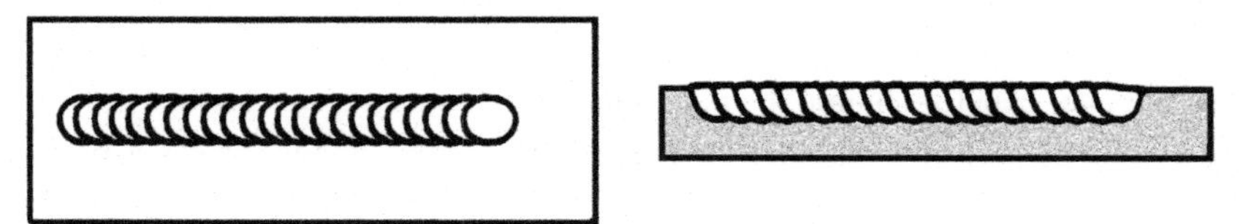

Figure 8.16 Thermal pulsing.

The pulse current is higher than would normally be used for continuous welding. Supplying heat in this way, at a high rate but for a short time, makes more efficient use of the energy. Benefits are better control of the weld pool and less sensitivity to gap width variations.

Combining thermal pulsing and high frequency pulsed drop transfer is used for aluminium welding. Using pulsed MIG during the high current period and the short arc settings during background time make it possible to weld very thin aluminium or stainless material. A high current spray arc combined with low current pulsed MIG is suitable for positional welding in thicker materials.

AC MIG welding

AC MIG welding is performed with special square wave power sources where the zero-crossing is fast enough to secure re-ignition of the arc. This type is often used for short arc welding or pulsed MIG welding of vehicle components in the lower range of current, up to 200 A.

Negative polarity on the electrode means a lower heat input and higher deposition rate. For AC MIG welding special power sources are used that produce a square-wave AC current. A balance control allows the user to adjust the ratio of negative to positive polarity. In practice that gives the user a tool to change proportion between heat and amount of filler material (see Figure 8.17). Advantages are that thinner material can be welded and it is possible to allow more variations of the gap width. Heat input and deformations decrease.

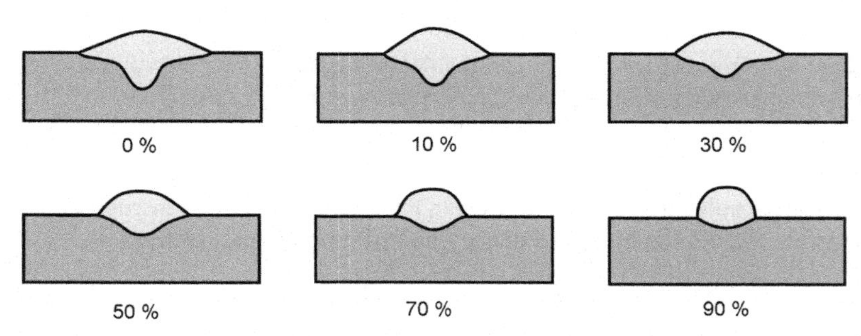

Figure 8.17 Penetration profile of AC MIG welding at different proportion of negative polarity.

Single-wire high productive MIG/MAG welding

The productivity of mechanised MIG/MAG welding, using conventional solid filler wires, has constantly improved in recent years. One of the leaders in this development was John Church, a Canadian who launched the TIME method (Transferred Ionized Molten Energy). Compared to conventional MAG welding, this method intentionally uses a long, high-current filler wire stickout. Resistive heating means that the wire is preheated, thus permitting a higher rate of feed without a corresponding increase in the current.

The AGA and Linde welding companies have investigated the method, and further developed it for use with other gas mixtures, calling the resulting processes Rapid Processing™ and Linfast® respectively.

The higher feed speed results in higher productivity: in some cases, at a rate of up to 20 kg/h of deposited weld metal (see Figure 8.18). The travel speed can be twice that of conventional MIG welding, while producing the same weld bead and penetration profile. Different types of arc are used: perhaps the commonest is a type of forced short arc that is within the range covered by conventional welding equipment.

Under certain conditions, a rotating arc is produced when welding at higher welding data. The high productivity, in combination with a higher current and larger weld pool, mean that welding must be carried out in the horizontal position.

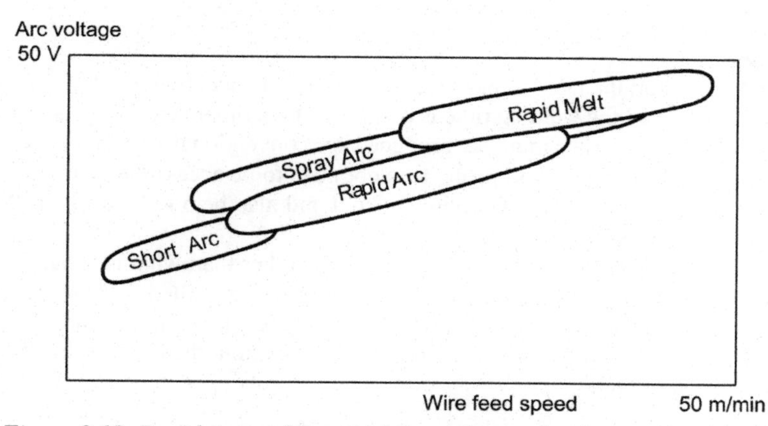

Figure 8.18 Rapid Arc and Rapid Melt can be used with considerably higher welding data than with conventional short arc or spray arc welding. (Royal Institute of Technology, Department of Welding, Stockholm, Sweden.)

Tandem and twin wire welding

Another way of improving productivity and raising the welding speed is to use double filler wires. Both wires can be connected to the same power unit, which means that they share a common arc. This method is referred to as twin arc welding (see Figure 8.19). Alternatively, if two power units are used, the method is referred to as tandem welding. The two wires are so close to each other they weld into a common weld pool.

Welding with two wires can increase the speed to at least twice the normal value or, when welding thin sheet, even higher. In some applications, linear welding speed can be up to 6 m/min.

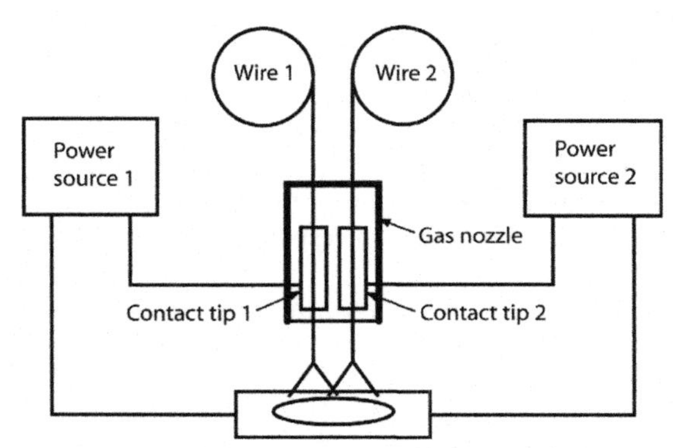

Figure 8.19 Tandem MIG welding uses two electrodes, normally positioned one after the other. Each electrode is separately connected to a power source.

Setting the welding current and voltage can be more complicated when two wires are used, particularly with tandem welding as it is necessary to set the welding data separately for each wire. Because the two arcs are close to each other, they can sometimes interfere through magnetic arc blow effect. Therefore this process often uses pulsed welding, with the pulses on each wire displaced out of phase with each other.

Arc spot welding

This is a MIG/MAG method intended to produce spot welds. The welding gun has a gas nozzle with support legs, and the welding time is controlled by a timer (see Figure 8.20). The joint type is often an overlap joint, as produced by conventional resistance spot welding. However, in this case, the workpiece does not need to be accessible from both sides. The support legs provide the correct wire stickout and also help to press the two pieces of metal into contact with each other.

Welding data (voltage and wire feed) are considerably higher than what is usual for the particular metal thickness concerned. The welding time is controlled to produce a fully penetrating weld within a relatively short time, generally less than one second. This produces a weld with low convexity and good coverage and without the need of drilling a hole in the upper piece of metal.

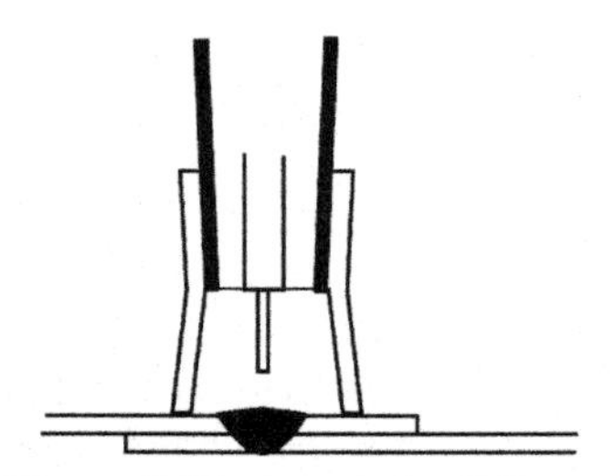

Figure 8.20 Arc spot welding.

In comparison with continuous welding, the process has the following advantages:

- Less heating and distortion.

- Very simple to operate: simply position and press.
- Lower, better-shaped convexity, particularly when welding thin sheet.

As the welding process is short but intensive, the method is less sensitive to welding position, imbalance in the metal thicknesses, gap width variations etc., and can therefore be used when such effects would otherwise make it difficult to produce successful continuous welds.

Welding parameters

The success of the MIG/MAG welding process is also dependent on a number of welding parameters:

- Voltage
- Wire size
- Wire feed speed and current
- Wire stick-out length
- Welding speed
- Inductance
- Choice of shielding gas and gas flow rate
- Torch and joint position

Most of these parameters must be matched to each other for optimum welding performance. The *working point* must be within the working range or *tolerance box* for the particular welding situation.

Voltage

Increased voltage increases the arc length and gives a wider weld bead. Undercut is a sign of too high a voltage. If short arc welding is used a higher voltage reduces the short circuit frequency, which will give larger drops and more spatter. Too low a voltage, on the other hand, will increase the risk of stubbing and problems in starting.

Short arc welding of thin plates allows a high welding speed without burn through. Normally the voltage here is adjusted to a low setting but only where the short circuit frequency is still high and the arc stability good.

Wire size

The size of wire is chosen according to the welding current. In contrast to covered electrodes, each wire has a large and overlapping range of current. As a rule, material transfer is smother with a thinner wire. When welding with soft aluminium wire, the risk of feeding problems can be reduced with a thicker wire. The price will also decrease with an increased wire size.

Wire feed speed and current

Current is set indirectly by the wire feed speed and diameter. It is the main parameter for welding and has to be chosen in relation to plate thickness and welding speed with respect to the weld quality.

Wire extension

In working out the wire extension, it is easiest to measure the contact tip distance from the joint surface, (see Figure 8.21). A rule of thumb is that a normal distance is 10–15 x diameter of the wire. Too small a stick-out increases the risk of burn-back, where the arc will weld the wire together with the contact tip. Too long a distance to the workpiece will increase the risk of stubbing, especially at the start.

The contact tip-to-work distance also has an influence on the current and penetration profile. If the wire extension is increased the current and heat input decreases while the amount of deposited metal remains. This reduces the penetration and if it was unintentional, will increase the risk of lack of fusion. A good rule is therefore to keep the wire stick-out constant during the welding operation.

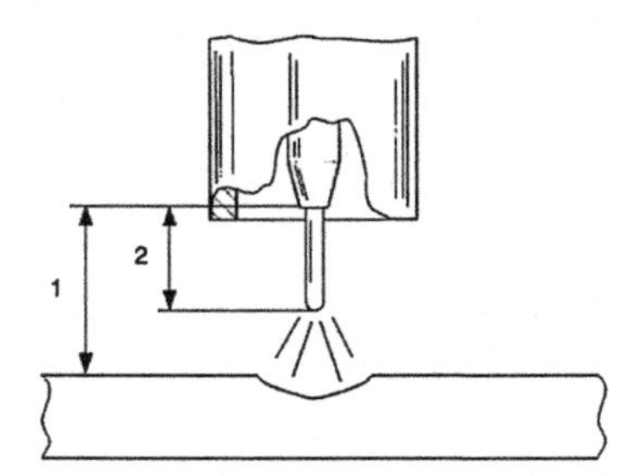

Figure 8.21 Definitions of contact tip-to-work distance (1) and wire extension (2).

Welding speed

Welding speed has also a considerable effect on shape and penetration of the weld, (see Figure 10.7 on page 113).

Inductance

It is often possible to adjust the inductance of the power source to fit the wire size and to give the right welding properties. The most sensitive process is short arc welding. A low value gives a distinct and concentrated arc but the spatter will increase. A higher value gives a wider bead and a softer sound. Too high an inductance reduces stability and increases the risk of stubbing.

Choice of shielding gas

Mixtures of argon with 5–20 % carbon dioxide (CO_2) and sometimes some oxygen (O_2) are most popular for the welding of mild and low alloyed steels. For spray and pulsed arc welding, a low content of CO_2 can be an advantage. A lower content of CO_2 and O_2 will also reduce the amount of slag on the surface of the weld bead. Pure CO_2 is an alternative for short arc welding that gives good penetration and safety against lack of fusion but increasing the amount of spatter. Argon is used for stainless steels with small additions (1–3 %) of CO_2 or O_2. In gases for duplex steels an addition of about 2 % nitrogen is recommended.

For welding of aluminium, copper and copper alloys pure argon or argon helium mixtures are normally used. Helium increases the heat input, which will compensate for the large heat conduction in thick walled aluminium or copper.

Gas flow rate

The gas flow must be adapted to the arc. At low current 10 litres per minute is sufficient while, at higher current up to 20 litres may be required. Welding in aluminium needs more gas than steel does.

The gas regulator reduces the pressure and keeps the gas flow at the correct value. In spite of this there will be a higher flow just at the start of welding and this gives increased gas consumption. The degree of the problem depends on the length and diameter of the hose between the regulator and the magnetic valve. A two-stage regulator has a lower outlet pressure and the build-up of gas will be less.

Torch and joint position

The angle of the torch relative to the joint is also an important welding parameter. If it is directed away from the finished part of the weld (*forehand technique*), it will make the penetration profile more shallow and the width of the seam wider (see Figure 8.22). On the other hand, if it is directed towards the finished part of the weld (*backhand technique*), the penetration will be deeper and the seam width narrower.

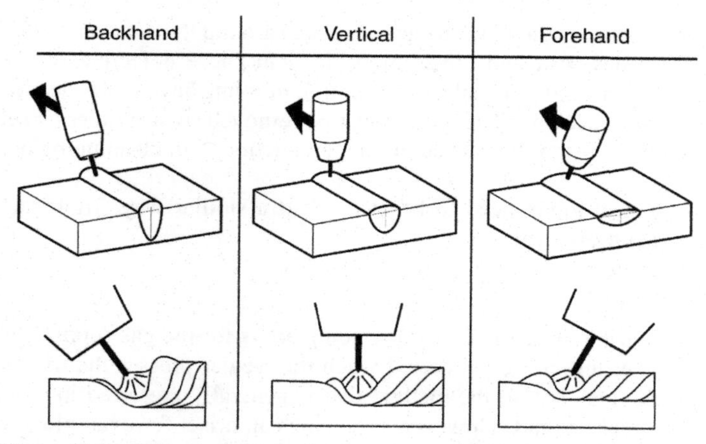

Figure 8.22 Effect of electrode position and welding technique.

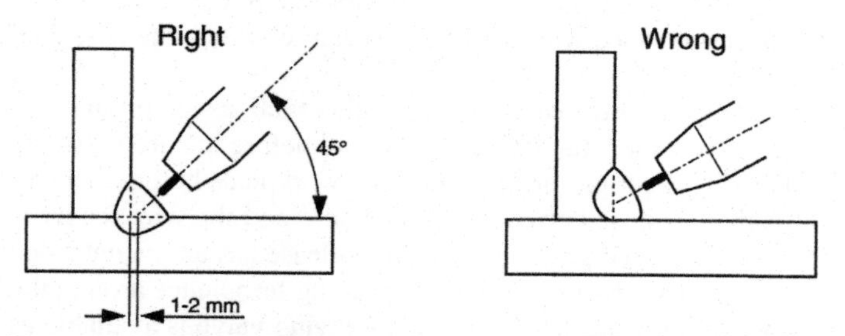

Figure 8.23 Gun angle and position across the welding direction at fillet welding. The wire is on such thick plates often positioned 1–2 mm offset on the base plate. That will compensate for the higher heat dissipation in the base plate and gives a symmetrical penetration profile.

The angle of the torch in the section across the welding direction has a direct effect influence on the risk of lack of fusion (see Figure 8.23). To reduce the risk of lack of fusion it is essential to prevent the melted metal from flowing too much before the arc. This can be the case when welding with high heat input, a large pool and welding with too much forehand or in downhill position. If the plate to be welded is not totally horizontal but has an inclined joint, it will also affect the weld contour and penetration profile. By welding downhill, the weld reinforcement can be lower and the welding speed will usually increase. At the same time, the penetration is lower and weld bead wider. This is beneficial for welding sheet metal. Uphill welding causes the weld pool to flow back and form a high and narrow weld.

8.5 Quality issues in MIG/MAG welding

There are a number of key operations that must be carried out properly if weld quality is to be assured.

Joint preparation

Before commencement of welding, the joint surfaces and area around the weld should be cleaned. Moisture, dirt, oxides, rust and other impurities can cause defective welds. Stainless steels and aluminium needs special care. Aluminium joints have to be cleaned by degreasing with alcohol or acetone. The oxide must be removed by a stainless steel brush or other ways. Welding must to be carried out in conjunction with cleaning before new oxide builds up.

High quality stainless steel joints need to be cleaned in a similar way. Tools and brushes must be made of stainless steel.

Starting welding

The normal procedure for striking the MIG/MAG welding arc is for the gas supply, the wire feed unit and the power unit all to be started when the welder presses the trigger switch on the welding gun. This is also the method that is generally preferred in most cases, as it results in the quickest start. However, problems can arise: for example, the wire may hit the joint, or there may be one or more false starts before a weld pool is created and the welding stabilises.

Creep starting provides a gentler start. The wire is fed forward at reduced wire feed speed until electrical contact is established with the workpiece, after which the wire feed speed increases to the set value.

Gas pre-flow is used when welding sensitive materials, such as aluminium or stainless steel. The gas flow starts a short (and adjustable) time before the arc is struck. The function ensures that there is proper gas protection of the workpiece before welding starts. Note, however, that if the gas hose between the gas bottle and the wire feeder is long, it can act as a 'store' for compressed gas, which is then released as an uncontrolled puff of gas when the gas valve opens, involving a risk of creating turbulence around the weld, with reduced protection from the gas. However, a gas-saving valve is available as an accessory, reducing the pressure from the gas bottle and thus eliminating the risk of a puff of gas. The gas pre-flow function will also eliminate this problem because the puff is rather short.

The *hot start* facility increases the wire feed speed and arc voltage for a controllable time during the start of welding. It reduces the risk of poor fusion at the start, before full heat inflow has become established.

Unstable starting is a problem that can occur from a number of causes. The tendency increases if the inductance is high and the voltage is low.

Post weld treatment

The corrosion resistance of stainless steels is degraded if oxides from welding remain. Special root gas is used to avoid oxidation and chemical and mechanical post weld treatment has to be undertaken.

Weld defects

It is important to be aware of common defects in MIG/MAG welding, particularly if the correct parameters are not maintained. Common defects are:

- Pores
- Lack of fusion
- End craters

Pores

Pores in the weld metal (see Figure 8.24) are often caused by some disturbance in the gas shield, but there are also other causes:

- Wrong setting of the gas flow. The flow must be adjusted according to the welding current. Too high a volume of gas will cause problems with turbulence in the gas nozzle.
- Draughts where the welding takes place. Airspeed above 0.5 m/s can interfere with the gas stream from the gun (depending on the setting of the gas flow).
- Defective equipment. Clogged channels or leakage can prevent the gas from flowing. Check, if possible, the gas flow by measuring directly on the opening of the gas nozzle. It is important to clean the inside of the nozzle regularly to remove spatter.
- Joint surfaces contaminated with oil, rust or paint.

Figure 8.24 X-ray picture of an aluminium weld containing pores.

Lack of fusion

Problems with lack of fusion between weld and parent metal (see Figure 8.25) have different causes:

- Incorrect setting of welding parameters. Low voltage or long wire stick-out will result in welding where the added heat is not in proportion to the amount of filler material.
- The melted metal in the pool flows before the arc. This can be caused if the weld is sloping, and if the welding gun is not properly aligned with the direction of the weld. The problem will also be worse if there is a large weld pool as a result of high heat input and slow travel speed of the welding gun. The arc must be kept on the leading edge of the pool and directed to prevent weld metal from rolling over at the front end.
- The large thermal dissipation of thick workpieces.
- Unfavourable geometry or too narrow a joint angle.
- The arc is directed in the wrong way (misalignment) and one edge of the joint is heated insufficiently.

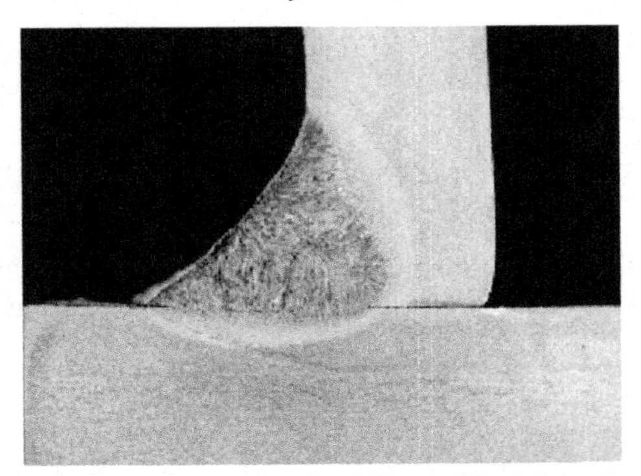

Figure 8.25 Example of a fillet weld with lack of fusion against the base plate.

End craters

End craters arise as a result of direct interruption in welding (Figure 8.26). This allows a crack or crater to form when the final part of the metal solidify, as a result of shrinkage forces during solidification. When grains from opposite sides grow together, low melting-point constituents and impurities can be swept ahead of the solidification front to form a line of weakness in the centre of the weld.

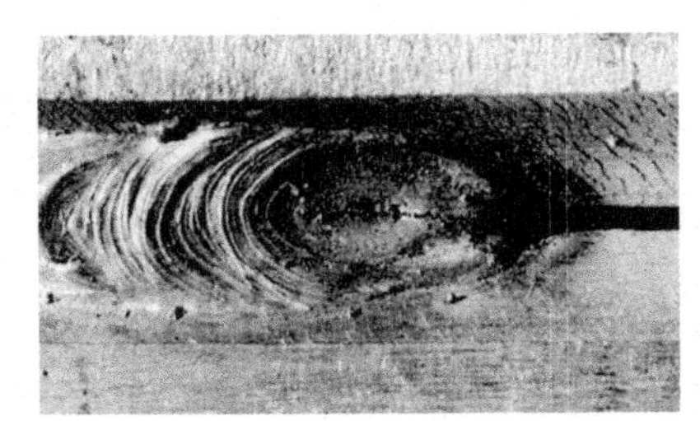

Figure 8.26 A crack may appear as an effect of direct interruption of the welding.

The *crater filling* function available in advanced power sources can be used to avoid the creation of craters when welds are finished. The arc continues to provide a reduced heat input while the weld pool solidifies. This has the effect of modifying the solidification process so that the final part of the weld pool solidifies at the top, thus avoiding the formation of a crater.

8.6 References and further reading

K. Weman and G. Lindén, *MIG welding guide*, Woodhead Publishing Limited, 2006.

ESAB and AGA, *Facts about MIG/MAG welding*. ESAB Welding Equipment AB.

V. Ponomarev, A. Slivinsky, *Welding arc and MIG/MAG welding transfer*, Kiev, 2003.

P. Houldcroft and R. John (1988), *Welding and cutting*, Woodhead Publishing Limited 1988.

9 Manual metal arc (MMA) welding with coated electrodes

9.1 Introduction

Manual Metal Arc welding (MMA) is often referred to as Shielded Metal Arc Welding (SMAW) or stick electrode welding. It was the predominant form of fusion welding until the beginning of the 1980s. It uses electrode rods consisting of a wire core with an external coating containing mixtures of substances such as chemicals, minerals and iron powder. They are made in a range of core diameters, with each diameter being intended for a particular current range. Welding involves striking an arc between the electrode and the workpiece, with the heat of the arc melting the electrode coating which forms a protective slag. The weld metal is produced both by the core electrode wire and iron powder in the coating. The layer of slag on top of the joint needs to be removed after welding.

The equipment required is simple, as shown in Figure 9.1, which means that the method is straightforward to use. It is particularly suitable for jobs such as the erection of structures. It can also be used outdoors, as opposed to other methods requiring shielding gas, which are unsuitable in wind. However, its arc time factor is relatively low, due to the time required for chipping away slag after welding and changing the electrodes.

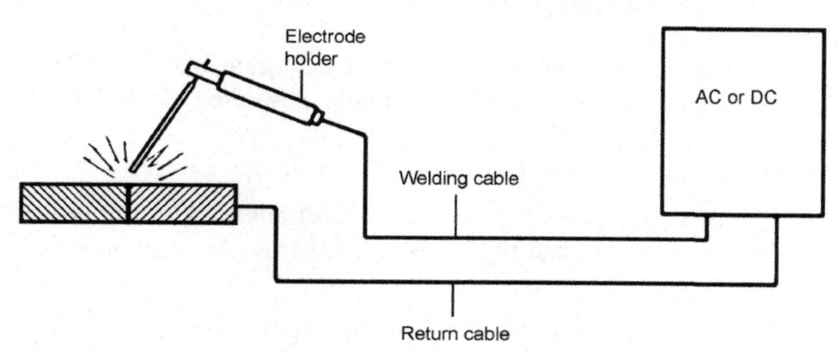

Figure 9.1 Schematic diagram of manual metal arc welding.

9.2 Equipment

When welding with coated electrodes, the required current is set at the power source. The welding current must be kept within certain limits even though the arc length may vary. The short-circuit current should not be more than about 60 % higher than the current setting, in order to avoid spatter from the short-circuiting of the arc caused by the droplets of molten metal in it. On the other hand, too low a rise in current can result in the electrode 'freezing' to the workpiece. These requirements mean that the power source should have a drooping load characteristic. Power sources for MMA welding do not need to have as high a duty cycle as those for more mechanised methods: a normal value is 35 %.

MMA welding can be performed using either DC or AC, which means that many types of power sources can be used. The advantages of using AC are that the power source is simple, deposition rate is high and there is reduced magnetic arc blow effect on the arc. However, AC-welding restricts the choice of type of electrode and require the power source to provide a sufficiently high open-circuit voltage, of at least 50 V or preferably more.

Mechanisation of MMA welding is possible using what is known as *gravity arc welding*, which involves securing the electrode and feeding it along the joint using a mechanical electrode feeder frame (Figure 9.2). In this way, a single welder can keep 3–5 arcs burning at the same time. With correctly set welding parameters, the use of this cheap and simple equipment enables one welder to make 280–400 m of fillet welds in an 8-hour working day. Such a productivity level is difficult to achieve with any other type of welding equipment in a similar price class.

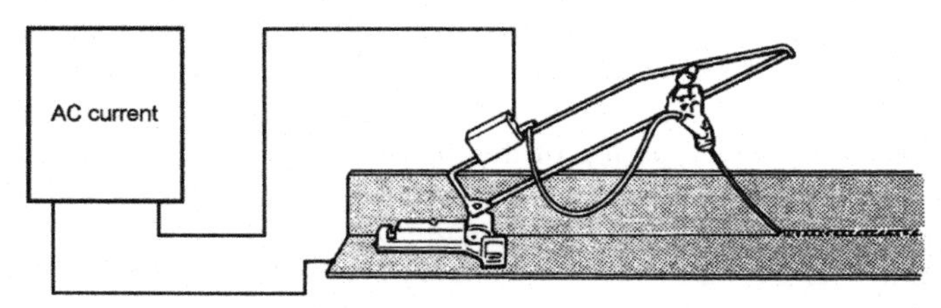

Figure 9.2 Gravity arc welding equipment for 700 mm long electrodes.

9.3 Consumables: electrodes

A wide selection of electrodes is available, to meet most requirements. The coating around the wire core consist of various mixtures of finely powdered chemicals and minerals, held together by a suitable binder.

The coating performs a number of important functions, including:

- *protecting* the metal droplets and the weld pool against reactions with the air; protection is provided by the molten slag and the gases emitted from the coating as it is heated.
- improving the *stability of the arc*. Without arc-stabilising substances in the coating, the arc would be difficult to control and would produce excessive spatter. It would also extinguish easily, particularly when welding with AC. Arc-stabilising or ionising substances include titanium, zirconium and magnesium.
- *shaping* the upper surface of the weld and facilitating *removal of the hardened slag*. The use of coated electrodes produces a layer of slag on top of the joint that has to be removed after welding. This requires the use of a chipping hammer or wire brush, and can be easier or harder, depending on the type of electrode coating.
- *applying alloying and/or anti-oxidising substances* to the weld pool. The coating may also contain iron powder in order to improve the efficiency (the yield).
- providing sufficient *penetration* into the base material while welding. Penetration is determined by materials that can release a significant quantity of hot gas, such as carbonates or cellulose compounds.

Electrodes are divided into three main groups, depending on the chemical composition of the slag: **acid**, **rutile** and **basic**. The factors influencing the choice of these three main types are summarised in Table 9.1. Other types of electrode which will be discussed in this section are cellulosic and penetration electrodes.

The efficiency of an electrode is calculated as the percentage of the weight of all weld metal from the electrode in proportion to the weight of the core wire. One can distinguish between normal efficiency electrodes, having a yield of up to about 130 %, and high efficiency electrodes, which have a yield in excess of 130 %, often up to 150–250 %.

Acid electrodes

The coating of acid electrodes includes iron and manganese oxides. The efficiency of an electrode can be substantially improved by mixing iron powder with the coating. The productivity of acid electrodes is good because of the high iron powder content. Acid electrodes produce smooth, shiny weld beads. The slag solidifies slowly, is porous and easy to remove. The weld metal has a lower yield strength and ultimate tensile strength than that produced by rutile and basic electrodes, but a higher rupture strain. Electrodes of this type show excellent welding characteristics and are suitable for welding in the horizontal position. Their principal application is the welding of unalloyed steel with low carbon content.

Rutile electrodes

The coating of rutile electrodes contains large quantities (about 25–45 %) of the mineral rutile (TiO_2). Electrodes of this type produce an arc that is easy to strike and re-strike. They are very easy to use and produce neat welds with an easily removable slag. The evenness of the weld bead and the resulting good connection to the base metal at the joint edges generally results in high fatigue strength. Unfortunately, these electrodes produce a higher hydrogen content in the weld metal, which introduces the risk of hydrogen embrittlement and cracking and restricts their use to welding carbon steel having a minimum ultimate tensile strength less than 440 MPa. Rutile electrodes are suitable for standard structural steel and shipbuilding steel.

Basic electrodes

Basic electrodes contain calcium fluoride (fluorspar – CaF_2) in the coating. The slag reacts as a base, thus leaving low sulphur and oxygen contents in the weld metal. The strength and toughness of the weld are therefore the strongest welds produced by any type of electrode, and the resistance to hot cracking is also higher. Basic electrodes produce a slag having a lower melting point than that from rutile and acid electrodes, which means that the risk of slag inclusions is slight, even if the slag has not been completely removed between passes. They are well suited for positional welding in all positions.

Due to the very high temperatures involved (up to 500°C) in the manufacture of basic electrodes, the moisture content of the coating is low when the electrodes are supplied. As a result, the hydrogen content in basic weld metals is low, thus providing good cold cracking performance. Basic electrodes are, however, hygroscopic, which means that they must be stored in dry conditions. When delivered, they are normally packed in special diffusion-proof wrappings. Every welder should have access to a container in which the electrodes can be kept at a temperature of 50–80°C throughout the working period. At the end of the day, the container can be emptied, and unused electrodes stored in an oven at a temperature of about 150°C.

A disadvantage of basic electrodes is that they produce a somewhat coarser and rougher weld surface, generally of a convex shape. Typical applications are structural, shipbuilding and pressure-vessel steel.

Cellulosic electrodes

The coating of cellulosic electrodes contains a relatively high proportion of cellulose substances, intended to produce excellent penetration by providing a high hydrogen content in the arc when welding in any position. The coatings have a high moisture content which means the resulting weld metal contains a high dissolved hydrogen content. It is therefore necessary to employ special methods of welding and to operate at elevated temperatures of 100–250°C even when welding relatively thin materials (>8–10 mm). The principal application of cellulosic electrodes is welding of oil or natural gas pipelines, using the vertical down position.

Penetration electrodes

Penetration electrodes are coated with a thick mixed rutile/acid coating, containing a high proportion of cellulose substances. They produce substantial gas emissions which increase the arc penetration in the underlying base material. These electrodes are used only in the horizontal position and for welding I-joints (square butt joints).

TABLE 9.1 Factors influencing the choice of welding electrode.

Factor	Rutile electrode	Acid electrode	Basic electrode
Arc stability	++	+	-
Appearance of the weld bead	+	++	-
Striking the arc	++	+	-
Strength of the weld	-	-	++
Different welding positions	+	--	++
Risk of slag inclusions	-	-	++
Resistance to corrosion	+	+	-
Fume formation	++	+	-
Slag removal	+	++	-
Hydrogen content in weld metal	-	-	++

9.4 Quality issues

There are a number of quality issues to consider in MMA welding. The most important are:

- Defects from slag inclusions in the weld
- Defects from poor penetration of the weld

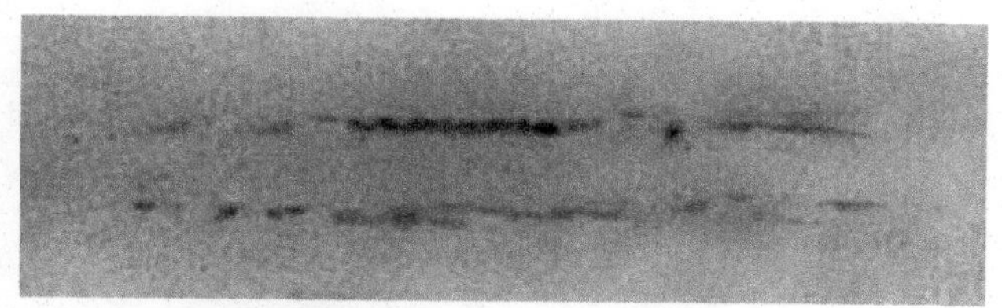

Figure 9.3 X-ray film showing slag inclusions.

Slag inclusions

As been noted, the use of coated electrodes produces a layer of slag on top of the joint that has to be removed after welding. Slag particles tend to get caught in hollows and sharp angles. Removing slag requires the use of a chipping hammer or wire brush, and can be easier or harder, depending on the type of electrode coating. If slag is not properly removed, it can lead to slag inclusions in the weld. The inclusions may be in the form of individual particles or longer lines (see Figure 9.3). The most common cause of slag inclusions is failure to completely remove the slag from one weld pass before making another. Good working methods can reduce the risks. It is also important to use the correct welding method and to avoid the use of excessively thick electrodes in confined joints. It is also important to try to weld in such a way as to avoid undercutting.

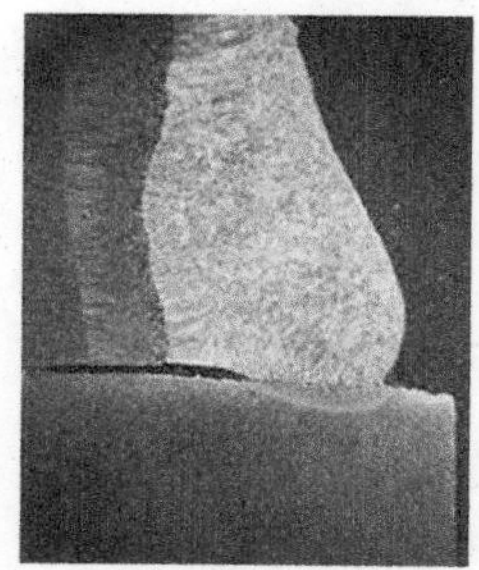

Figure 9.4 Poor penetration gives a root defect.

Poor penetration

A common form of root defect or poor penetration is that caused by insufficient penetration or the presence of slag residues in the root. It can be particularly difficult to achieve full penetration when restarting after replacing an electrode, and conditions are made more difficult by an uneven gap or joint shape. In such cases, it is often necessary to grind or chip the root side of the weld and to re-weld.

9.5 References and further reading

J. Houldcroft and R. John, *Welding and cutting*, Woodhead Publishing Limited, 2001.

10 Submerged arc welding

10.1 Introduction

Submerged arc welding (SAW) (Figure 10.1), is a high-productivity method of welding, generally carried out using mechanical welding methods and suitable for use with 1–3 continuous wire electrodes.

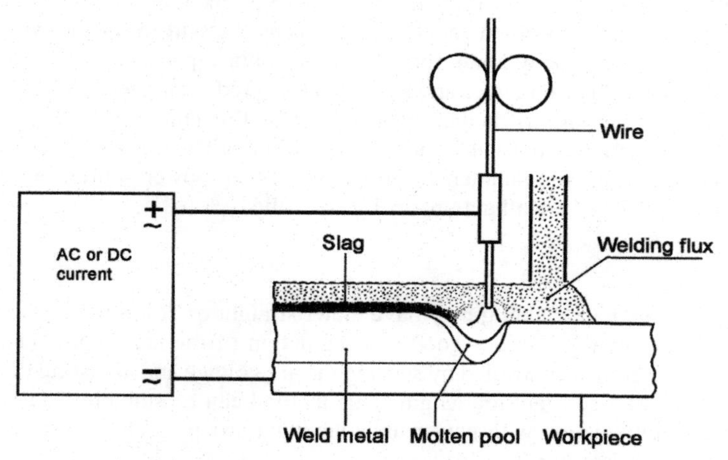

Figure 10.1 Schematic diagram of submerged arc welding.

The arc or arcs are struck and burn beneath a layer of flux, which is supplied to the welding head whilst welding is in progress. The flux closest to the arc melts and forms slag on the surface of the weld, thus protecting the molten metal from reacting with the oxygen and nitrogen in the air. Residual powder is sucked up, returned to the flux hopper and re-used. Welding can be carried out with DC or AC.

If the welding parameters are properly set, the appearance of the weld is often very uniform and bright, merging smoothly into the workpiece material. The slag also usually comes away by itself. The flux masks the light from the arc and there is no smoke or spatter from the weld. This improves working conditions as compared to that of gas metal-arc welding. On the other hand, there is still the need to handle the flux. Although it is supplied automatically to the weld and recovered as part of the process, the use and handling of flux is still a complicating factor.

The advantages of the submerged arc welding method are:

- A high deposition rate
- Deep penetration, which allows the quantity of filler material to be reduced
- The ability to achieve a high arc time factor (i.e. effective welding time)
- High weld quality
- An improved working environment compared to other arc welding methods.

Submerged arc welding is suitable for welding sheet metal from about 1.5 mm thick and upwards, although it is mostly used for somewhat thicker materials. The process is used mainly for large items such as plates in shipyards, longitudinal welding of large tubes or beams, or large cylindrical vessels. The method is used for both butt welds and fillet welds in the horizontal position. Another application area is cladding, e.g. of stainless steel onto ordinary mild steel, or when applying a coating of some hard wear-resistant material. These processes often use strip electrodes.

10.2 Equipment

Welding equipment for submerged arc welding consists of a wire feed unit, in the form of a drive motor, reduction gear and feed rollers, which feeds the wire from the wire spool to a contact device, preferably with spring-loaded contact pads. The flux is supplied to the weld from a flux container, and is often recovered after the weld by a suction unit which sucks up the surplus flux and returns it to the flux container.

The movement of the torch is normally mechanised, although there are welding torches intended for semi automatic submerged arc welding. The power source, wire feed speed and linear travel speed are all automatically controlled.

Power sources

Power sources for submerged arc welding may have either straight or drooping characteristics. A straight characteristic provides good self-regulation of the arc length. The wire feed speed is sometimes also controlled by sensing the arc voltage and adjusting the wire feed speed to maintain a constant arc length. This method can be suitable for use with thick wires and in combination with power units having a drooping characteristic, in order to reduce current variations.

Power sources for submerged arc welding are designed for high current and duty cycles, e.g. 800–1600 A, 44 V and 100 % duty cycle. Both AC and DC welding may be used. One type of AC welding current power source produces a square wave. This is a relatively simple and satisfactory way of controlling single-phase AC without interrupting the welding current and extinguishing the arc. If the welding cables are long, it is a good idea to run the supply and return cables close together. This reduces the inductive voltage drop, assists the fast zero transitions and avoids the unnecessary creation of magnetic fields around the cables.

Arc striking methods

It can sometimes be difficult to strike the arc if the power source has a low short-circuit current (drooping characteristic). Other causes can be slag on the electrode wire, or flux between the wire and the workpiece. There are many ways of assisting striking, of which the most common is to cut the wire, preferably to produce a sharp point.

Mechanisation

Equipment used for the longitudinal travel motion includes

- A welding tractor that runs directly on the sheet to be welded (Figure 10.2);
- A welding head that can be mounted on a column and boom unit (Figure 10.3);

- Powered rollers for rotating cylindrical workpieces (Figure 10.4).

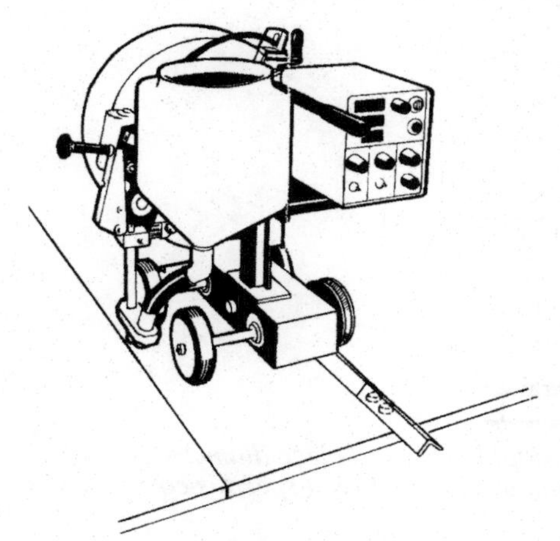

Figure 10.2 A welding tractor for mechanised welding.

It may also be necessary to have some kind of equipment to guide travel along the joint. One simple method is to project a spot of light in front of the welding point, and for the operator to keep this centred on the line of the joint. Another method involves purely mechanical control, using support rollers etc. In the case of larger workpieces, it may be appropriate to have some type of automatic joint tracking control. A common principle is to have a sensor finger that rides in the joint ahead of the arc, to provide servo control of a crosshead that carries the welding head.

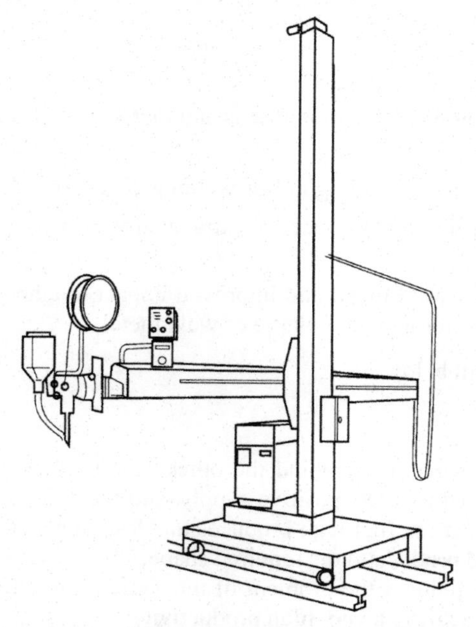

Figure 10.3 Welding head fitted to a column and boom unit.

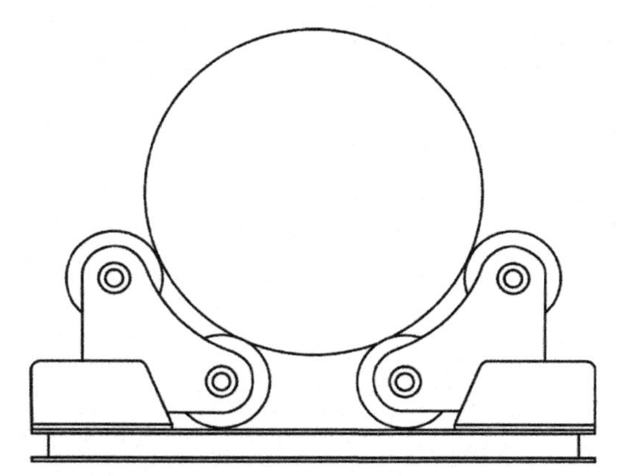

Figure 10.4 Roller beds that automatically adapt to the workpiece diameter. They are often used together with a column and boom unit. The rotational speed is controlled by the built-in motor.

Productivity improvements in submerged arc welding

There are a range of ways of improving the productivity of submerged arc welding. These include:

- Tandem welding
- Twin arc welding
- Cold wire welding
- Hot wire welding
- Welding with long stickout
- Welding with a metal powder additive

The impact of some of these techniques on the deposition rate is summarised in Figure 10.5. Other benefits of these methods are:

- Lower heat input for the same melt volume, resulting in less welding distortion.
- Control of the arc pool composition through the use of extra materials and less melting of the base metal.
- A narrower HAZ, finer structure of the weld metal and improved impact toughness as a result of the heat input being lower for a given volume of weld metal.

These methods are discussed in more detail below.

Tandem welding

This arrangement employs two or more wires, one behind the other, feeding into the same weld pool. The first wire is connected to a DC positive supply, and the second is connected to an AC supply. This means that the first wire produces the desired penetration, while the other(s) fill(s) the weld and produce(s) the required shape. The use of AC on the second and subsequent wires also reduces the problem of magnetic blow effect between the wires. This arrangement can achieve a very high productivity.

Twin arc welding

This involves feeding two wires in parallel through the same contact tip. It differs from tandem welding in using only one power unit and one wire feeder. Depending on the desired result, the wires may be arranged side by side or one behind the other. In comparison with the use of a single wire, twin arc welding results in a higher rate of melt production and improved stability.

A twin-arc welding machine can be easily produced by fitting a single-wire machine with feed rollers and contact tips for two wires. Without very much higher capital costs, it is possible to increase the deposition rate by 30–40 % in comparison with that of a single-wire machine. Wire sizes normally used for butt welding are 2.0, 2.5 and 3.0 mm, with wire separations of about 8 mm.

Long stickout

Increasing the distance between the point where the current enters the wire (the contact tip in the wire feeder unit) and the arc has the effect of resistance-heating the wire, giving a 20–50 % higher deposition rate, which means that the welding speed can be increased. However, the wire must be carefully guided: there is a risk of root defects if it is not properly aligned in the weld. A wire straightener can be the answer.

Cold wire

Two wires are used, but only one carries current: the other is fed into the arc from the side. This increases deposition rate by 35–70 %.

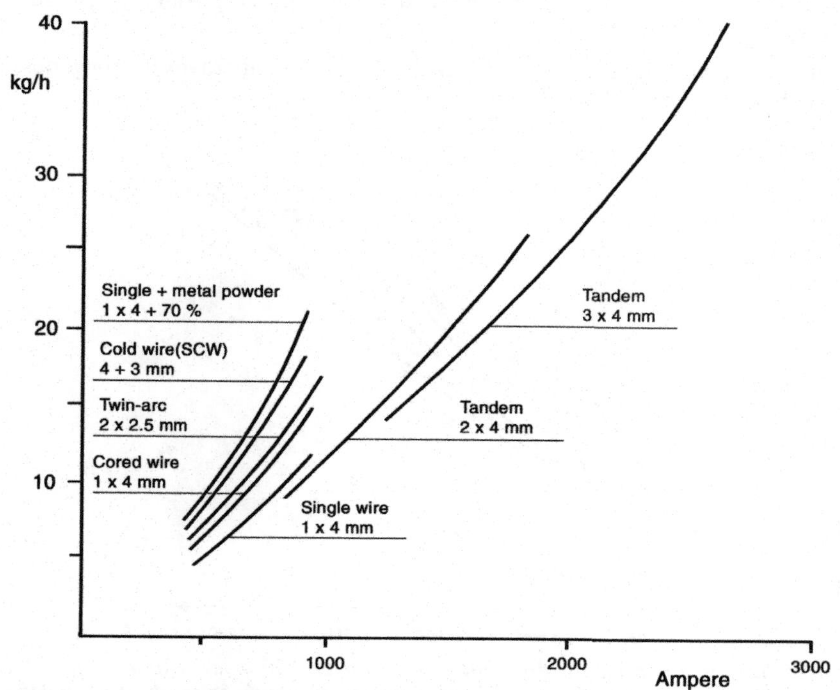

Figure 10.5 Illustration of different possibilities to increase the deposition rate.

Hot wire

This involves the use of an additional wire, resistively heated by the welding current. It can increase the quantity of melted material by 50–100 %.

Metal powder additive

Metal powder or small pieces of cut filler wire are fed into the weld, and melted by the heat of the arc. This can increase deposition rate by up to 100 %.

10.3 Consumables

The proper choices of filler wire and flux composition are important for the finished weld. The aim is to achieve a composition and strength of the weld metal similar to that of the base material. It is important to allow for such factors as possible loss of alloying elements by burn-off in the arc, melting of the base metal and alloying from the flux.

When using a strongly alloying flux in a joint with many passes, there is a risk of build-up of alloying material through uptake of material from previous passes.

Filler wires

The wire grade and its content of alloying metals primarily affect the mechanical properties and chemical analysis of the weld metal. When deciding on an appropriate choice of wire, it is very important to allow for the following factors:

- The strength of the weld metal can be increased by alloying with manganese and silicon.
- The use of molybdenum and nickel as alloying elements improves the toughness of the weld metal at low temperatures.

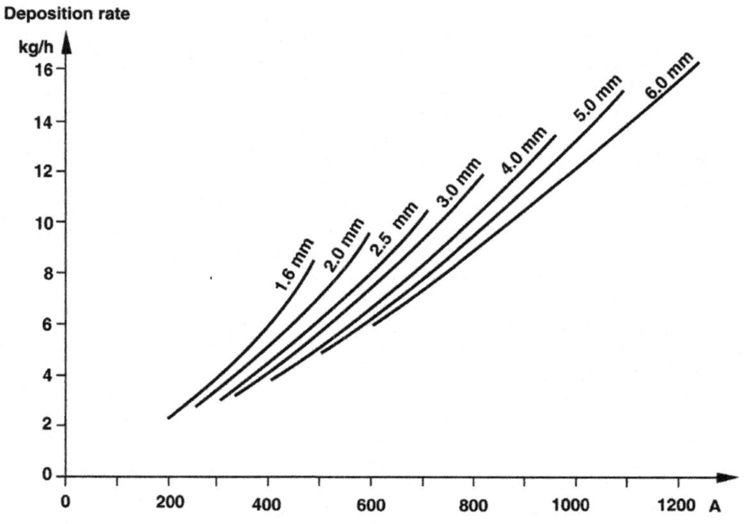

Figure 10.6 Deposition rates for different wire diameters. 30 mm stickout length, DC+ polarity. The higher melting rate for thinner diameters depends on the higher resistive pre-heating in the stickout.

The wire may be copper-plated in order to improve electrical contact and to protect against corrosion. Common wire diameters are 1.6, 2.0, 2.5, 3, 4, 5 and 6 mm (see Figure 10.6).

Filler material in the form of strip (e.g. 0.5 x 100 mm) is often used when applying stainless steel cladding, e.g. to pressure vessel steel. As a result of the wide cross-section, penetration is exceptionally low, producing a smooth and wide weld. The low dilution from the parent metal does not affect the corrosion resistance of the surface layer. The method is also used for repair of worn parts.

Flux

The most important purposes of the flux are:

- To form a slag and protect the molten weld metal against the harmful effects of the air.
- To supply alloying constituents to the weld metal and control its composition.
- To improve the stability of the arc and to assist ignition.
- To form the weld's surface convexity and give a good surface finish to the weld metal.
- To control the flow characteristics of the molten weld metal.

TABLE 10.1 The melting point of the flux has a considerable effect on the quantity of micro-slags in the weld metal. In this respect, the oxygen content is a measure of the quantity of these slag inclusions.

Type of flux	Basicity	Melt interval, °C	Oxygen content by weight, ppm
Acid	< 0.9	1100–1300	> 750
Neutral	0.9–1.1	1300–1500	550–750
Basic	1.2–2.0	> 1500	300–550
High basic	> 2.0	> 1500	< 300

As with coated electrodes, the flux may be acidic, rutile or basic. Acid and rutile fluxes have excellent welding characteristics and produce a good weld appearance, although the mechanical properties of the weld are somewhat lower. A high proportion of oxides in the form of microslags has an unfavourable effect on the impact toughness of the weld (see Table 10.1). Flux with high basicity improves the mechanical properties, but at the expense of somewhat lower welding performance. The best results are obtained if the depth of the flux is controlled when welding so that the arc is just hidden by the flux bed.

Flux manufacture

There are two main types of flux: fused flux and agglomerated flux. Table 10.2 summarises the main characteristics of each type

Fused flux. The preparation of fused flux involves melting minerals at high temperature to produce a glass-like mass. This is allowed to cool, and is then crushed and screened to appropriate grain sizes. Chemical homogeneity of the molten flux is important. It is also important that the flux should not be hygroscopic since this means less risk of damp flux. However, reactions can occur between the alloying substances when the flux melts, which imposes some limits on the chemical composition of the material.

Agglomerated flux. This is made by adding a suitable binder, such as water-glass, to the dry powder constituents of the flux. It is then dried in rotary kiln at a temperature of 600–900 °C, after which the material is screened to produce the required grain size. A characteristic of agglomerated flux is that it is easy to vary the chemical composition by adding various alloying elements, and that it is very hygroscopic. Agglomerated flux may also be more tolerant of rust and mill scale. Its popularity has increased so that it is now the most commonly used type of flux.

TABLE 10.2 Properties of fused and agglomerated flux.

Flux type	Benefits	Drawbacks
Fused	Non-hygroscopic High grain strength	Alloying elements such as Cr and Ni cannot be incorporated in the flux High density (approx. 1.6 kg/l)
Agglomerated	Alloying elements such as Cr and Ni can be included in the flux Low density (approx. 0.8 kg/l)	Hygroscopic Relatively low grain strength

10.4 Process knowledge

A number of factors affect the quality of welds in submerged arc welding, including joint preparation and the control of differing welding parameters.

Joint preparation, design and root support

As the submerged arc welding process is a mechanised method, the quality of joint preparation prior to welding is important. Insufficient attention to tolerances or cleaning will result in a defective weld. A clean, properly prepared joint also allows higher welding speeds to be used, with reduced repairing costs, which more than compensates for more expensive preparation.

Single-sided welding usually involves some form of root support:

- A backing tongue of steel, which is allowed to remain after welding.
- A backing support in the form of a water-cooled copper bar.
- A flux bed in a grooved copper bar.
- A special ceramic backing support.

Double-sided welding means that there is no need to provide root support. The good penetration of submerged arc welding means that butt welds can be made in plate up to 15 mm thick without requiring a gap or joint preparation. Thicker materials require joint preparation in the form of V-shaped or X-shaped joint faces, perhaps also in conjunction with multiple weld passes. Asymmetrical X-joint faces are used in order to even out distortion. The first pass is made in the smaller of the two gaps. As the weld metal cools and contracts, it pulls the plate slightly upwards along the line of the joint.

Other types of joint include fillet joints, which are very common. Narrow gap welding is preferable for welding very thick materials (see Page 160).

Welding parameters and weld quality

As with other arc welding methods, welding parameters have a considerable effect on the characteristics and quality of the welded joint. The variables in submerged arc welding are:

- welding speed
- polarity
- arc voltage
- arc current
- the size and shape of the welding wire
- the filler wire angle
- the number of welding wires
- wire stickout length
- the use of additional filler wire or metal powder additive
- the type of flux (acid / neutral / basic).

The welding speed affects the penetration and the width of the weld (the cross-sectional area of the weld) (see Figure 10.7). A high speed produces a narrow weld with little penetration. An excessively high speed produces a risk of undercutting, pores, root defects, poor fusion and magnetic blow effect. Too low a speed results in an uneven surface, while extremely low speed produces a mushroom-shaped penetration, and can result in solidification cracks. In addition, too low a speed produces a large weld pool which flows round the arc and results in an uneven surface and slag inclusions.

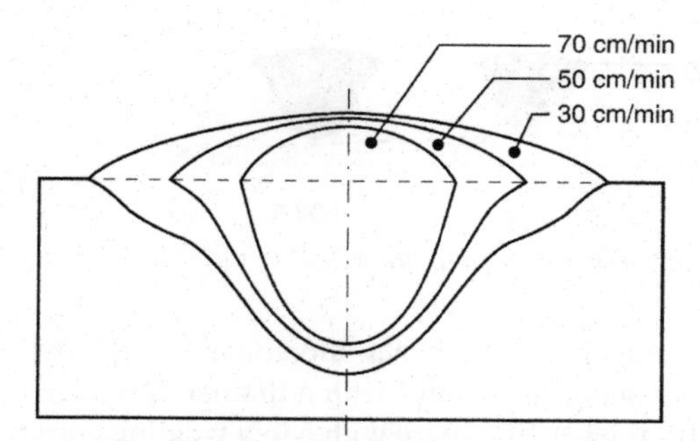

Figure 10.7 The effect of welding speed on weld appearance, with constant values of current and voltage.

Polarity also affects weld penetration (see Figure 10.8). If the filler wire is positive, penetration is deeper than if the filler wire is negative. This means that it is better to use negative polarity when performing cladding, in order to avoid mixing the cladding material into the base material. The melting rate is increased by about 30 % percent when negative polarity is used.

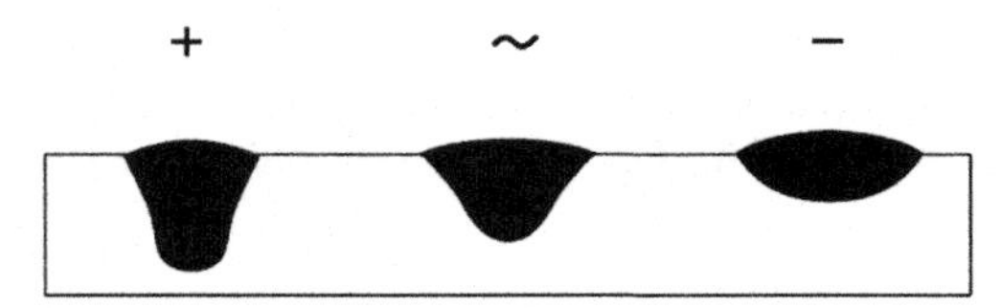

Figure 10.8 The effect of wire polarity on penetration.

A high **arc voltage** produces a broad weld with little penetration (see Figure 10.9). This means that it is suitable for welding wide gaps and for increasing the admixture of alloying elements from the flux. It also increases flux consumption and makes removal of the cold slag more difficult. A high arc voltage also increases the risk of undercutting, particularly when making fillet welds. A low arc voltage, on the other hand, produces a high weld convexity and an unsuitable merging angle with the base material.

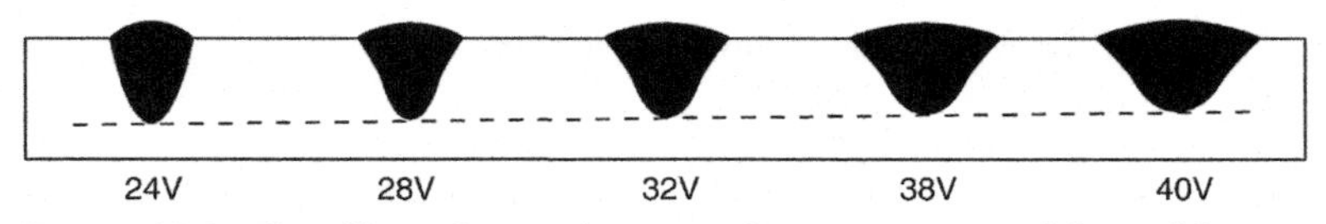

Figure 10.9 The effect of arc voltage on the appearance of the weld.

The **welding current** affects penetration and deposition rate as shown in Figure 10.10. However, too high a welding current can result in undercuts, an uneven weld convexity, burn-through, thermal cracking, an inappropriate merging angle with the base material and undercutting.

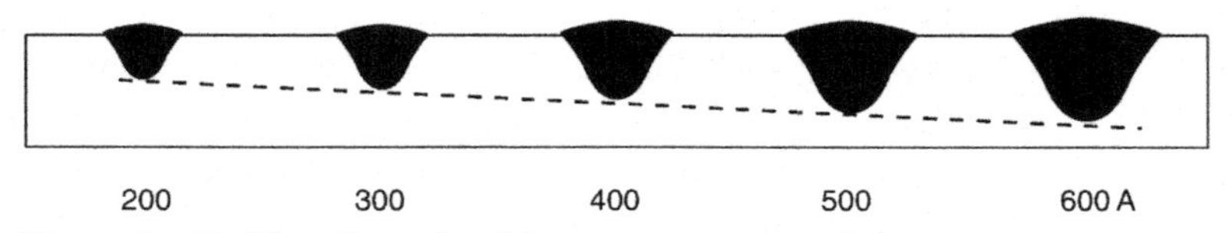

Figure 10.10 The effect of welding current on weld appearance.

Wire size. A smaller wire diameter results in greater penetration than a thicker wire (see Figure 10.11). For a given current, arc stability is better with a thinner wire, due to higher current density. On the other hand, a thicker filler wire with a low welding current can more easily bridge a wide joint.

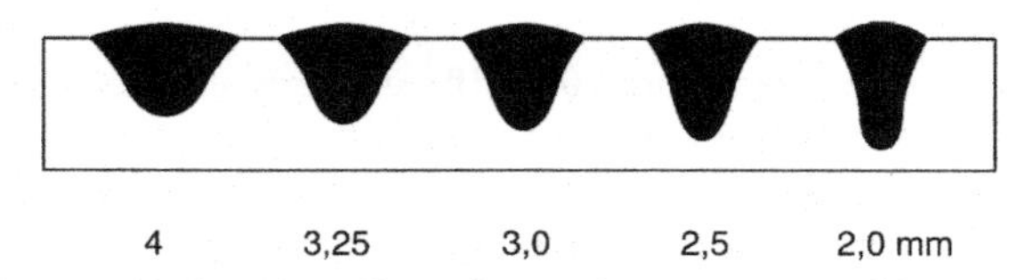

Figure 10.11 The effect of wire diameter on weld appearance.

Wire angle. The angle of the filler wire to the joint has a considerable effect on the shape and penetration of the weld (see Figure 10.12). In certain cases, forehand welding gives a wider bead that can counteract the tendency to produce a high, narrow weld convexity, and thus allow a higher welding speed to be used.

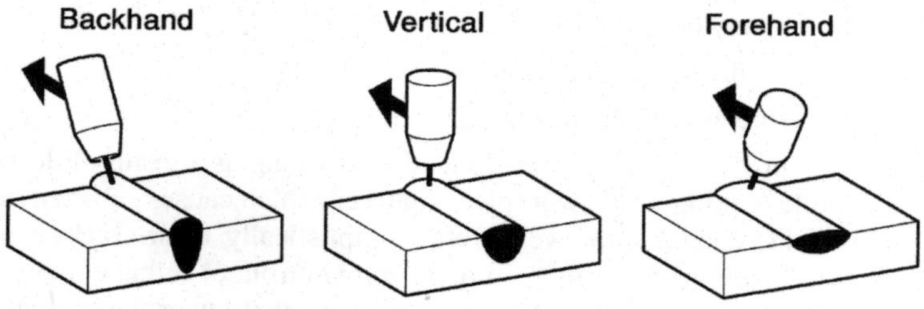

Figure 10.12 The effect of wire angle.

10.5 Quality issues: weld defects

There are a range of weld defects that affect submerged arc welding. These include:

- Hydrogen embrittlement
- Solidification cracking
- Pores and pinholes
- Poor impact strength
- Undercutting
- Slag inclusions
- Uneven weld beads

Hydrogen embrittlement

Hydrogen embrittlement is also referred to as *hydrogen cracking* or *cold cracking.* The cracks occur in the HAZ, close to the melt boundary as the material cools, sometimes several hours after welding. The effect is caused by a combination of shrinkage stresses, hydrogen diffusing in from the weld metal and the formation of the hard martensite phase structure.

A drawback of all welding processes involving protection by flux is the risk of moisture absorption and the resulting increased risk of hydrogen embrittlement. The flux should be properly stored in order to keep it dry. Materials having high carbon equivalents and thicknesses must be welded at elevated temperatures, in accordance with the relevant rules.

Hydrogen is introduced from the molten pool through moisture or hydrogen containing elements on the surface of the parent metal. The hydrogen diffuses from the weld bead to the adjacent regions of the heat affected zone. Fast cooling in combination with steels with higher strength can give a hardening effect. If hydrogen is present there is a great risk for hydrogen cracking. Thick plates and a low heat input gives a high cooling rate and this increases the risk of hydrogen embrittlement. An increased operating temperature of the workpiece and carefully dried consumables are an important ways to assure weld quality.

Solidification cracking

Solidification cracking, also called *hot cracking* arise as the material cools, if certain combinations of unfavourable conditions occur (see Figure 10.13).

- Low width/depth ratio of the weld penetration.
- High carbon and sulphur contents in the metal.
- Shrinkage stresses occurring as the material cools.

Submerged arc welding produces a risk of solidification cracking as a result of deep penetration and considerable melting of the workpiece material which causes substances from the workpiece material to end up in the weld metal. Simplistically, these cracks can be explained by the solidification front pushing a molten zone in front of it that contains higher concentrations of easily melted substances (or substances that lower the melting point) than in the rest of the weld metal. In a deep, narrow joint, the weld metal solidifies in such a way as to leave a weakened stretch trapped in the middle of the weld, which then breaks to produce a longitudinal crack under the influence of shrinkage stresses.

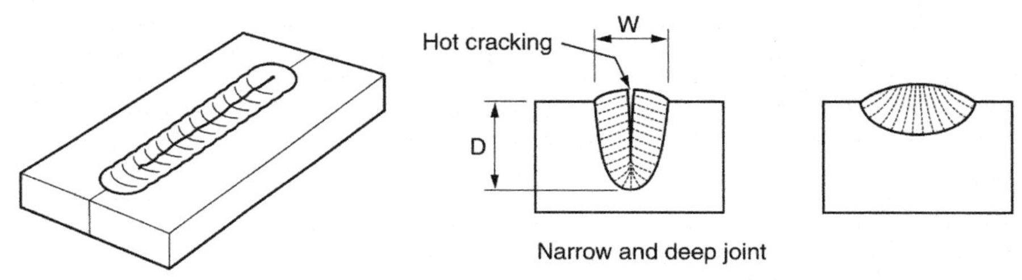

Figure 10.13 Solidification cracks may appear when the weld is deep and narrow.

Solidification cracking can be eliminated by forcing the weld to cool from the bottom towards the surface, so that the primary crystals are forced to grow diagonally upwards towards the surface of the weld, e.g. by welding against a heat-removing base.

Many highly alloyed steels have a wider range of temperature over which solidification takes place. This increases the susceptibility to solidification cracking. It is also strongly influenced by the solidification direction.

Welding defects associated with starting and stopping welding can be avoided through the use of starting and stopping tabs that are later removed.

Pores and pinholes

Pores can be caused by several factors, such as moisture in the flux and dirt on the workpiece, such as rust or paint. Problems with pores have a tendency to increase if the molten metal cools rapidly.

Pinholes are due to the release of gas (mainly hydrogen) during solidification of the metal, i.e. during primary crystallisation. The gas is unable to escape sufficiently easily from the weld metal. As a result, it is retained in the metal and acts as nuclei around which the metal solidifies. Pinholes form in the middle of the weld, running along it like a string of beads. Pinhole formation can be reduced by carefully cleaning the surface of the weld joint prior to welding and by reducing the speed of welding.

Poor impact strength

Poor impact strength due to grain growth occurs in connection with slow cooling. The high performance and good penetration of submerged arc welding means that it is best to weld even somewhat thicker materials with as few passes as possible. However, this results in high yield energy, so it may be better to make several passes when welding difficult materials.

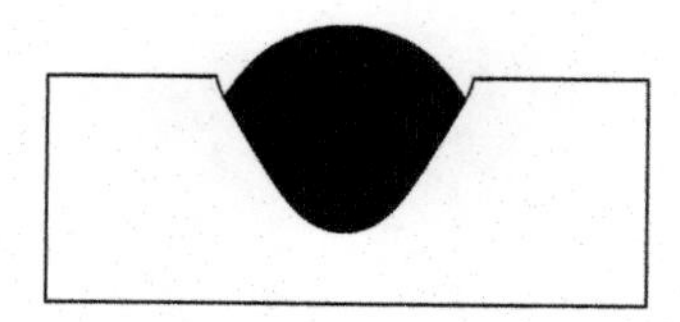

Figure 10.14 Too high voltage or welding speed may result in undercuts.

Undercutting

Undercutting is a defect that indicates that the appropriate voltage range for the process has been exceeded (see Figure 10.14). Too low a voltage results in a narrow, high weld convexity. Increasing the voltage makes the weld wider, but too high a voltage can easily cause undercutting at the edge of the weld bead. Too high a linear speed along the weld can cause both a high convexity and undercutting together. An undercut will appear when the weld metal doesn't fill up the cavity that is cut by the arc. It is most often troublesome in connection with welding of upright fillet joints, where it occurs in the web. If the angle of the wire is changed to a more forehand position, it may decrease the problem.

Slag inclusions

Slag inclusions are uncommon in SAW welds. If they do occur, it is usually between the passes in multi-pass welds. When making such welds in thick plate, care must be taken to remove all traces of slag.

Uneven weld beads

Uneven weld bead will result if welding current is high in comparison to the wire diameter – about 1100 A or more. This is caused by the high arc pressure on the weld pool. The result of this effect is that the penetration is excessive for the wire size in use, causing the molten metal to be ejected over the edge of the joint and sometimes also causing lack of fusion.

If the current for a wire is in excess of the recommended value, it is necessary to change up to the next wire size.

10.6 References and further reading

P. T. Houldcroft, *Submerged arc welding*, Woodhead Publishing Limited, 1990.

11 Pressure welding methods

11.1 Introduction

Pressure welding can be carried out by several welding methods which have in common the fact that the surfaces of the joint are pressed or worked together. Smiths in traditional forges used forge welding, which involved heating metal pieces in a fire until they became plastic and then forged them together. This means that forge welding could be classified among methods of pressure welding. In some cases (e.g. spot welding), the surfaces to be joined are heated to melting point, while in other methods the weld can be made without significant heating.

This chapter discusses the following pressure welding methods:

- Resistance welding (including spot welding, seam welding, projection welding, resistance butt welding and flash welding)
- Friction welding (including friction welding by rotation, friction surfacing and friction stir welding)
- High-frequency welding
- Ultrasonic welding
- Explosion welding
- Magnetic pulse welding
- Cold pressure welding
- Diffusion welding

11.2 Resistance welding

Resistance welding is one of the oldest types of welding technique. The different methods are generally fast, efficient and low-polluting. No filler materials are required. The drawbacks can be high capital cost and a somewhat limited range of applications. Each type of resistant welding method can generally only be used for one type of welding. Resistance welding equipment is also relatively expensive. As a result, the proportion of total cost accounted in relation to equipment cost is much higher than with arc welding.

In the process heat is generated by the passage of an electric current through the resistance formed by the contact between two metal surfaces. The current density is so high that a local pool of molten metal is formed, joining the two parts. The current is often in the range 1 000–100 000 A, and the voltage in the range 1–30 V.

To achieve joining, resistance welding machines must go through three basic steps:

1. Clamping or pressing together the workpieces with a certain mechanical force and holding them in the correct position.
2. Passing the necessary current through the workpiece.
3. Controlling the welding time as required.

There are two different types of machines, depending on the arrangement of the electrode arms: *swinging arm machines*, in which the upper arm is carried by a bearing in the frame, and *guide rail* machines, in which the upper electrode is controlled linearly by a pneumatic cylinder, as shown in Figure 11.1.

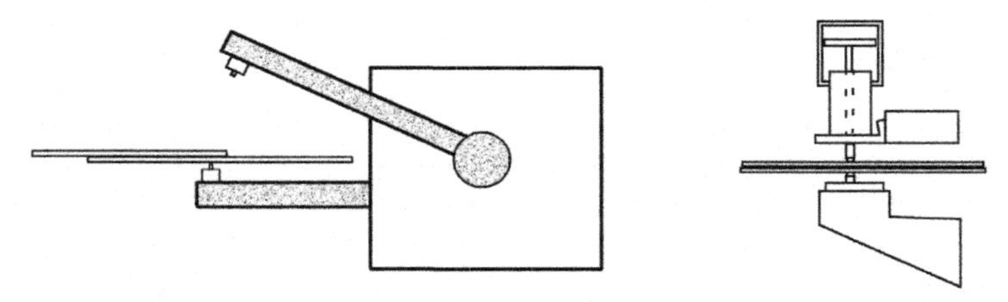

Figure 11.1 Swing arm and guide rail resistance welding machines.

It is important that the electrode arms are capable of moving quickly to accommodate the movement as the workpiece softens in the heat and moves together: if not, there is a risk of spatter from the weld. A mechanical or pneumatic spring can maintain a pressure on the electrode as the material 'collapses', thus reducing the risk of splatter.

The size of a machine and the length of the projecting arms depends primarily on the size and shape of the items to be welded. With AC welding the arms should not be longer than necessary, bearing in mind the electrical reactance of the loop enclosed by the arms, i.e. the area enclosed by the arms and the frame. (This applies, of course, only when welding with alternating current.) A large window area makes it possible to weld larger items, but also increases the reactive impedance. For this reason, the arms on most resistance welding machines are adjustable, although this does not apply for projection welding.

A tap-changer on the welding transformer provides basic (or coarse) control of the voltage and current. Fine control is then provided by the thyristor contactor which controls the switching of the welding current.

DC welding

DC welding machines, with a rectifier on the secondary side of the transformer, are more expensive but are immune to inductive voltage drop. They are also suitable for three-phase supplies, which provides a more balanced load on the mains and allows higher powers to be taken. It is nowadays also common to provide a DC supply using a medium-frequency inverter. The principle for this is the same as that for inverters used for arc welding: see Page 56. This reduces the size of the transformer and provides more rapid control of the current and so better control of the welding process. Electrode wear is also somewhat reduced. Resistance welding uses a medium/high frequency of about 1–4 kHz. Higher frequency (10–20 kHz) can be used to further reduce the weight on portable spot welding guns. As a DC welding machine does not suffer from reactive voltage drop, the total power demand from the mains is reduced.

The use of inverter technology in combination with intelligent technology in the power source permits exact real-time control of welding current and time, to deliver a better overall result.

If the pressure unit is controlled by servo motors, rather than pneumatically, cycle times can be reduced, e.g. in robot welding operations.

TABLE 11.1 Examples of applications for a number of resistance welding methods.

Item	Welding method			
	Spot	Projec-tion	Seam	Flash
Stainless steel sinks	•		•	
Wire meshes, storage trays etc.		•		
Furniture parts, chairs, tables		•		
Pipes, sleeves, nipples		•		
Tools, drills				•
Lockers	•			
Tops and bottoms of tanks			•	
Vehicle bodies	•			
Differential casings				•
Silencers	•		•	
Pipes and sections				•
Rails				•
Chain				•
Substantial girders	•			

We usually distinguish between five different types of resistance welding:

- spot welding
- seam welding
- projection welding
- resistance butt welding
- flash welding

Spot welding

Spot welding is the best-known resistance welding method. It is used for joining thin sheet materials (up to 3 + 3 mm) by overlap joints, and is widely used, e.g. in the automotive industry. A typical car can have up to 5 000 spot-welded joints.

The high current, in combination with a rapid heating time, means that the thermal energy input is efficiently used: very little is conducted away to the surrounding metal. Spot welding therefore has several advantages over other methods of welding sheet metal, such as:

- Little deformation of the workpiece, as the thermal energy is more or less restricted to the immediate vicinity of the weld.
- Very high rate of production for mechanised processes. Spot welding of 1 + 1 mm sheet, for example, takes 0.20 s.

- Easy to automate, with high consistency, making the method suitable for mass production.
- Low energy requirement and little pollution with no filler materials required. The technique therefore has less environmental impact than when welding with an arc.
- Little training required.

Two electrodes clamp the two sheets of metal together with a considerable force, while passing a high current through the metal. Thermal energy is produced as the current passes the electrical contact resistance between the two sheets, as given by:

$$Q = I^2 \cdot R \cdot t$$

where Q = quantity of thermal energy (Ws)
I = current (A)
R = the resistance across the weld (Ω)
t = welding time duration (s)

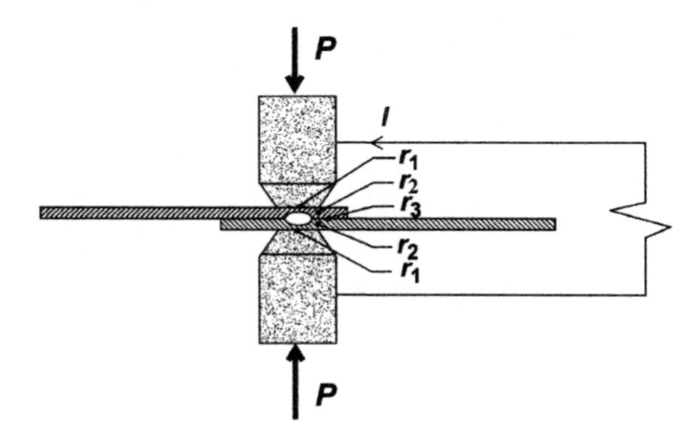

Figure 11.2 The principle of spot welding.

The total resistance between the electrodes (see Figure 11.2) is made up of:

$$2\,r_1 + 2\,r_2 + r_3$$

where r_1 = contact resistance between each electrode and the workpiece
r_2 = the resistance through the metal of each of the pieces to be joined
r_3 = the contact resistance between the two pieces of metal.

The contact resistance between the electrodes and the workpiece, and particularly the contact resistance between the two pieces of metal to be joined, is considerably higher than the resistance of the conducting path through the metal. Minor unevennesses in the surface of the metal means that the current is concentrated to a few contact points, with the result that the heating is greatest at the contact surfaces. Changing the clamping force can modify the contact resistance and thus also the heating of the metal.

As welding starts, the contact resistances are very high. The initial passage of current breaks through the surface layers, so that the contact resistances drop rapidly. Most of the heat formed at the contact between the electrodes and the workpiece is conducted away through the water-cooled electrodes. However, this is not the case with the heat developed in the contact resistance between the two workpiece sheets. The temperature here rises until the melting temperature of the metal is reached, while the surfaces continue to be pressed together by the clamping force, so that a weld nugget forms in the contact area.

The electrodes need to be of a material with a high hardness, low electrical resistance and high thermal conductivity. Cooling is decisive for their service life. Wear and tear, together with deformation, increase the effective contact size of the electrodes, which reduces the current density and, in due course, the strength of the welds produced. An electrode normally has a life of about 5 000–10 000 welds: when welding galvanised steel, this life is reduced to about 500–2 000 welds. Tip dressing, using a special tool, restores the shape of the electrode tip.

The spot welding process includes a number of parameters or variables that can be adjusted in order to achieve optimum welding performance. Tables of optimum values have been produced, but it is also necessary to optimise the process by trial and error.

The welding current is the current that flows through the workpiece. Of all the parameters, this has the greatest effect on the strength and quality of the weld, as the amount of heat produced is proportional to the square of the welding current. The welding current must therefore be carefully adjusted: too high a current results in a weld with poor strength, with too great a crater depression, spatter and some distortion. It also means that the electrodes are worn unnecessarily. Too low a current, on the other hand, also produces a weld of limited strength, but this time with too small a weld area.

Squeeze time is the time needed to build up the clamping force. It varies with the thickness of the metal and with the closeness of the fit, and is also affected by the design of the electrode jaws.

The clamping force is the force with which the electrodes press the sheets together (kN). It is important that this should be carefully controlled, as too low a clamping force results in a high contact resistance, accompanied by spatter and resulting in a poor weld strength, while too high a force results in too small a weld, again with poor strength, but accompanied by unnecessary wear on the electrodes and too great a crater depression.

Welding time is the time for which current flows through the workpiece, and is measured in cycles, i.e. the time for alternating current to pass through one cycle. In Europe, the mains frequency is 50 Hz, which means that one cycle takes 1/50 = 0.02 s.

Hold time is the time from when the current is interrupted until the clamping force can be released. The plates must be held together until the weld pool has solidified, so that the joint can be moved or the electrodes moved to the next welding position.

The electrode area determines the size of the area through which the welding current passes, i.e. the current density. The electrode diameter (d) is determined in relation to the thickness of the metal (t) from the following formula:

$$d = 5 \cdot \sqrt{t}$$

The welding parameters may need to be adjusted when welding high-strength steels, in order to avoid any risk of causing microcracks or pores.

The area in the diagram (see Figure 11.3) within which an acceptable spot weld can be produced is referred to as the tolerance box or weldability lobe. Too high a current results in spatter, while too low a current, or too short a welding time, results in an inadequate, or even no, weld nugget.

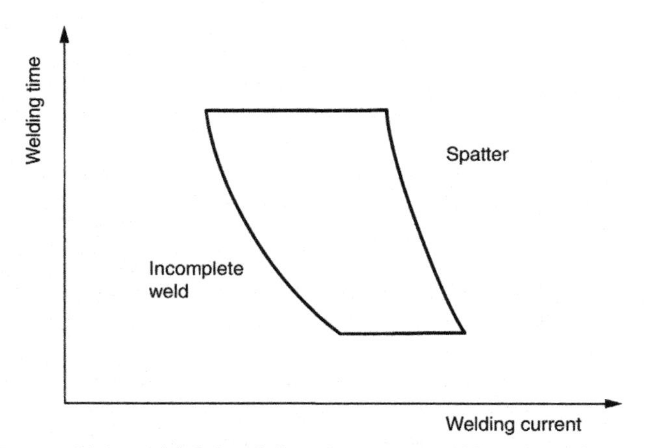

Figure 11.3 Weldability lobe where acceptable spot welding can be obtained.

Seam welding

Seam welding is used in the same way as spot welding, and operates on essentially the same principle. The difference is that two wheel-shaped electrodes are used, rolling along (and usually feeding) the workpiece (see Figure 11.4).

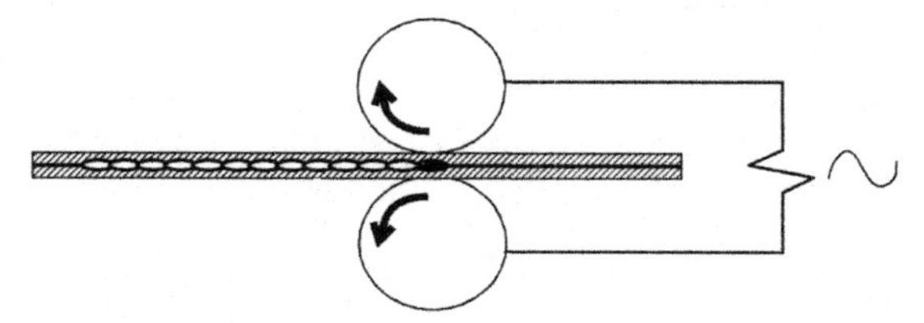

Figure 11.4 Principle of seam welding.

The two wheels should be of the same size, in order to prevent the part from being deflected towards one of them. The actual contact profile can be designed in a number of ways, in order to suit the shape of the part to be welded. The current may flow continuously while welding is being carried out, or intermittently to produce a series of spots that are so closely positioned as to produce a single, continuous weld. An unavoidable problem of seam welding is that some of the current 'leaks' through the completed weld.

As the electrode rollers rotate, they do not need to be lifted between each spot, as with spot welding. If the weld does not have to be continuous, seam welding can be used to position spots at regular distances from each other. That means that seam welding can be carried out quicker than ordinary spot welding.

Projection welding

As with seam welding and spot welding, projection welding is used to join two overlapping sheets of relatively thin metal. The process involves pressing a number of 'dimples' in one of the plates and welding the two plates together at the same time (see Figure 11.5).

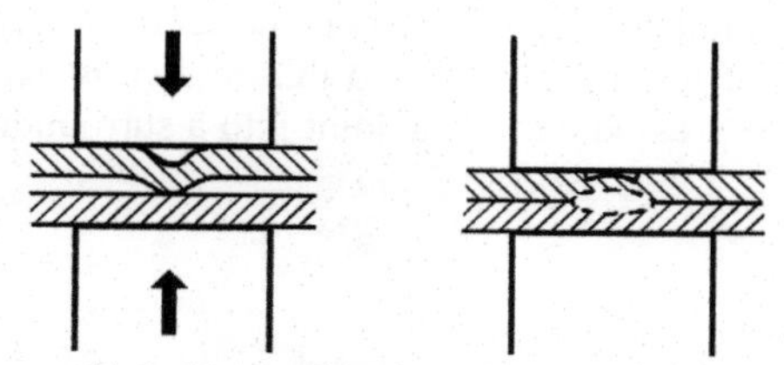

Figure 11.5 Principle of projection welding.

The method can also be used for welding metal sheet to the ends of bars, rods or pipes, or for welding nuts to sheets. Wire grids (i.e. the crossing points of the wires) are also particularly suitable for projection welding.

An advantage of the process, relative to spot welding, is that there is less wear and tear on the electrodes due to the greater contact area.

Resistance butt welding

Resistance butt welding is used for end-to-end welding of rods or wires, e.g. when welding wire baskets, shopping trolleys or wire racks for use in ovens. Butt welding can be used for welding steel, copper, aluminium and its alloys, as well as for gold, silver and zinc.

The ends of the material are pressed together and a current is passed through them (see Figure 11.6). The temperature across the contact resistance becomes so high that the metal softens to a plastic state and the two parts can be joined together. The maximum contact area is usually stated to be about 150 mm^2. The upper limit is determined by the ability of the welding machine to ensure even distribution of the heat across all parts of the joint. The lower limit is determined by the practicalities of handling the material: for steel wire, the smallest size is generally regarded as being about 0.2 mm diameter.

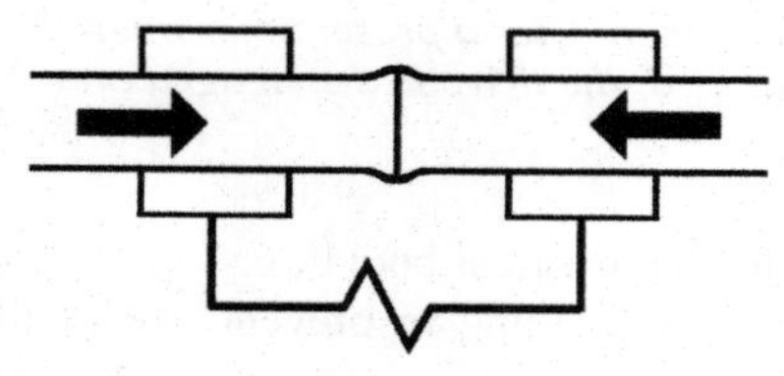

Figure 11.6 Resistance butt welding.

Flash welding

As with butt welding, flash welding is a method in which the ends of the workpiece are pressed together and welded. It is used for welding thicker workpieces such as heavy anchor chain, rails and pipes. The process is most often used for welding of steel but also for nickel, aluminium and titanium alloys.

The process starts by *preheating* the components. This is done by moving the parts forwards and backwards, into and out of contact with each other a number of times, while current is passing. When the temperature is sufficiently high, the process move on to the next stage, known as *flashing.* The parts are slowly brought together and pressed firmly in contact, which causes rapid melting and gasification, with spectacular ejection of molten material in a rain of sparks. The molten metal of the two surfaces joins, and the process continues with the application of forging pressure so that molten material and any trapped oxides or contamination are pressed out of the joint into a surrounding collar or upset.

11.3 Friction welding

Friction welding does not involve complete melting of the joint surfaces. The surfaces are heated up and affected in various ways by pressure and friction. The energy input is purely mechanical. The method has been used for more than 30 years. Traditionally, the necessary friction has been generated by relative movement between the workpiece parts, although in recent years the technology has been further developed so that the necessary friction-generating motion can be applied by an external tool.

Friction welding by rotation of one part of the workpiece

This is the traditional type of friction welding technique. It is very suitable for certain applications, particularly where at least one of the parts is rotationally symmetrical. This part is rotated and pressed against the other, producing a weld through simultaneous heat generation and plastic deformation, as shown in Figure 11.7. Parameters of importance are the speed of rotation, the pressure and the time for which the part is rotated.

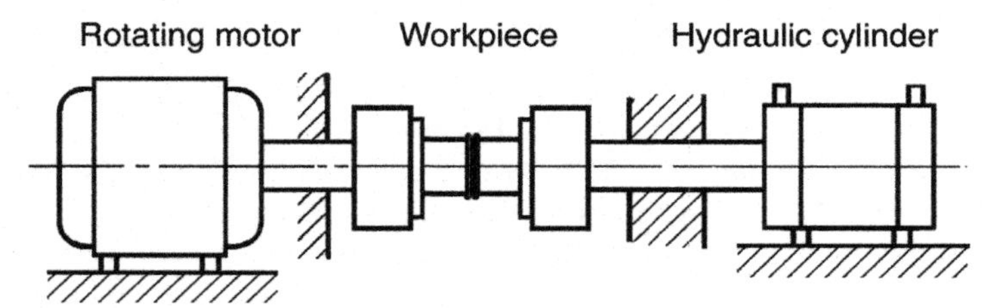

Figure 11.7 Friction welding, with the motor driving one part of the workpiece at a controlled speed while the two parts are pressed together. After a preset time, when the necessary amount of frictional heat has been developed, the drive is disengaged and rotation stops.

The method is usually used for two rotationally symmetrical bodies, e.g. bar to bar, pipe to pipe, flanges or endpieces. However, there is nothing to prevent one of the surfaces from being flat.

The welding process produces a collar of material that is pressed out of the joint, removing any surface contamination so that the joint is homogenous and free of defects (see Figure 11.8).

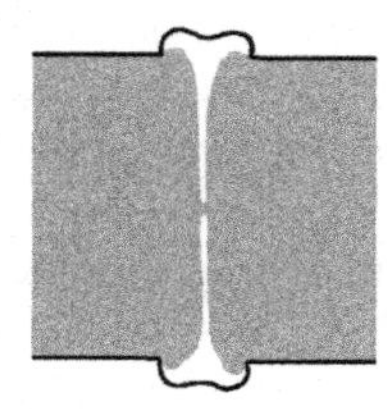

Figure 11.8 Section through a friction-welded bar.

Friction surfacing

Surfaces can be cladded using the friction welding principle. This involves a round bar of consumable materials that is rotated while being pressed against, and moved over, the workpiece surface. The process is very suitable for welding different combinations of materials: austenitic stainless steel can, for example, be applied to ordinary carbon steel.

Trials of this method have also been used for making welded joints. One method involves applying filler material into a continuous joint between two sheets of not too thick material. If the material is thicker, holes can be drilled in it which can then be filled by a rotating filler bar.

Friction stir welding

Friction stir welding (FSW) is an interesting development of earlier friction welding methods. It has also been developed as friction stir spot welding (FSSW), i.e. spot welding of overlapping plates with a FSW tool.

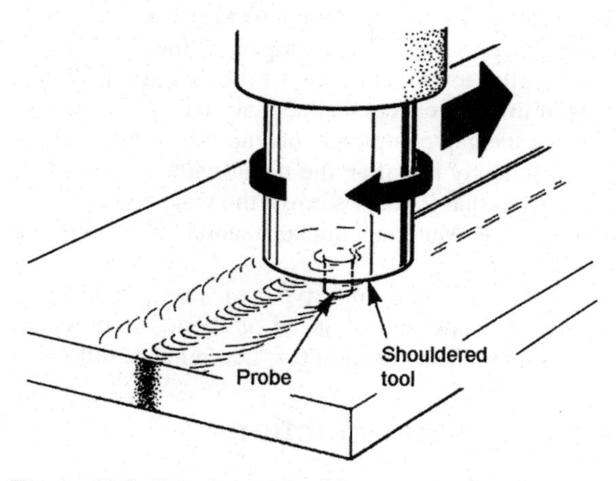

Figure 11.9 Friction stir welding.

The process is particularly suitable for welding aluminium, e.g. for making longitudinal welds along aluminium extrusions. It is also possible to use the method with certain other materials such as copper, titanium, lead, zinc and magnesium. Trials of welding plastics have also been carried out. The advantages of the method are as follows:

- The quality of the joint is consistently good. The root face can be so good that the weld is almost invisible, while the top is essentially smooth, but with a puddled surface effect left by the rotating tool.
- The welded joint has excellent fatigue strength.
- With a low heat input, there is very little thermal stress or distortion.
- Mechanical properties are better preserved compared to arc welding. A hardness drop of just 10–20 % has been measured.
- No joint preparation is necessary.
- FSW may be used also for alloys that are crack sensitive when they are welded with normal fusion welding processes.

- There is no visible radiation, noise or fume generation.
- No filler materials are required.
- The production rate is comparable with that of other methods.
- The method shows good profitability due to very little need for preparation or subsequent processing.

As well as a number of advantages, the process also has some drawbacks, including the following:

- The formation of a hole from the tool where it stops can be a disadvantage. The problem of producing an invisible termination has not been solved.
- Heavy and powerful fixtures are needed to keep the parts of the workpiece together and pressed to the backing plate.

The process uses a rotating tool, with a pin that penetrates almost completely through the workpiece (see Figure 11.9). The joint is a gap-free butt joint, and requires no special joint preparation. The workpiece must, however, be firmly clamped, as high pressure is generated as the tool passes. This also applies to the root face if full weld penetration is required. The method is similar to that of milling, except that no material is cut: instead, it is pressed past the rotating pin and fills the gap completely behind it. Friction and the 'stirring' effect raise the temperature sufficiently to soften the metal without melting it. The shape of the rotating tool is designed so that it presses down the weld convexity so that it remains level with the original surface. Materials thicker than about 25 mm are most often welded from both sides.

The welding speed depends on the thickness and type of material: 15 mm aluminium, for example, can be welded at a longitudinal speed of about 180 mm/min, while 5 mm thick material can be welded at a linear speed of 0.8–2 m/min (Al 6082).

11.4 High-frequency welding and induction welding

This method of welding could be regarded as a form of resistance welding, as the heat is created by resistive heating of a current induced in the workpiece. The use of a very high frequency, e.g. 400 kHz, concentrates the current close to the surface of the material, known as the skin effect, or in parts of the workpiece close to a current-carrying conductor, known as the proximity effect. This provides a means by which heating can be restricted to, or concentrated in, those parts of the surfaces to be welded together, with the final weld being made by pressing the parts together.

Current can be supplied to the workpiece by contact blocks or sliding shoes. The fast, concentrated heating provides a high rate of welding, with low heat input and little conduction of heat to other parts. An interesting application is that of longitudinal seam welding of pipes, for which welding rates of 30–100 m/min can be achieved, depending on the thickness of the material and the power input.

Induction welding is a form of high-frequency welding in which the current is induced in the workpiece by a coil, through which the high-frequency current flows, positioned close to or around the workpiece.

11.5 Ultrasonic welding

Ultrasonic welding bonds the workpiece parts together by vibrating them against each other at high frequency under pressure. To some extent, the equipment used for this is similar to that used for resistance welding, except that it is vibration, rather than electric current, that provides the energy input to the workpiece. Supported against a solid base, the parts to be welded are pressed together by the tool that transfers the vibration. An electromechanical ultrasonic head operates at a frequency in the range 20–60 kHz, with an amplitude of only a few hundredths of a millimetre. A layer of oxide, or even of insulation, need not prevent a good connection. However, the surfaces should be thoroughly degreased, as grease acts as a lubricant and degrades the quality.

Ultrasonic welding is suitable both for welding of plastics and metals. Ultrasonic welding of metal is used for thin sheet, films or wires. Parts to be joined should preferably be fairly small: at least one of them – the one that will be made to vibrate – should not be more than a few millimetres in size. An example of the use of this method is that of making electrical connections in aluminium or copper. Very thin conductors can also be welded in this way. The particular advantage of this method when using it to weld items in sensitive electronic equipment is that it produces very little heat.

11.6 Explosion welding

Explosion welding develops an extremely high pressure for a short time. The surfaces to be joined are brought together at very high speed. The impact energy plasticises them and produces a good welded bond (see Figure 11.10). The amount of heat developed is modest. As opposed to other processes, such as those involving melting of the materials, there is no (or only an insignificant) melt zone in which materials from the two pieces can be mixed or can chemically react with each other. Explosion welding of larger pieces must be performed at specially designated sites.

Explosion welding is used primarily for preparing blanks or billets, although it can also be used for such applications as joining pipes or securing tubes in tubeplates. The method is often used for a combination of materials that are difficult to join using other methods. An example of such an application is the bonding of aluminium sections to steel, on which it is then possible to use ordinary welding methods to add further aluminium components, e.g. the superstructure of a vessel, made of aluminium, on a steel hull. Another application is for the manufacture of compound plate, in which a base plate is plated with a sheet or plate of another metal, as shown in Figure 11.10.

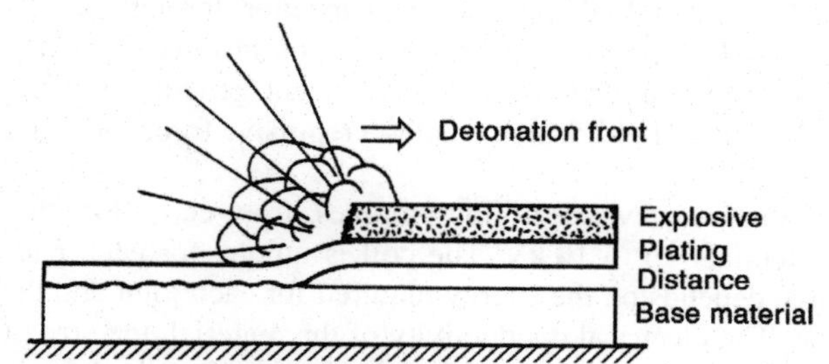

Figure 11.10 Applying plating to a billet by explosion welding. (Nitro Metall AB)

Figure 11.11 Micro photograph of an explosive-welded joint in unalloyed steel. (Nitro Metall AB)

11.7 Magnetic pulse welding

Magnetic pulse welding (MPW) is a more recent welding technique. Its closest relative is explosion welding but, instead of an explosive, an extremely powerful magnetic pulse is used. In comparison with explosion welding, important advantages include a low noise level and an ability to make repeated welds with short time intervals.

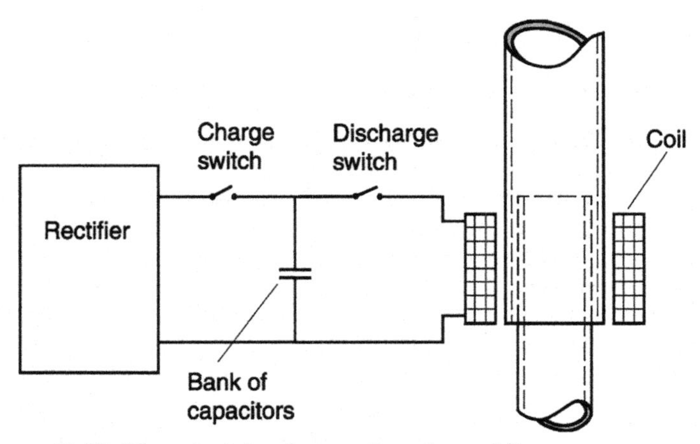

Figure 11.12 The principle of magnetic pulse welding.

The principle involves discharging a high-capacitance capacitor through a coil surrounding the parts to be welded. Two pipes, for example, can be inserted into each other to form an overlap joint (see Figure 11.12). The high current generates a brief (<1 ms) but extremely powerful magnetic field, which creates a repulsive force, pressing the outer part towards the inner one.

The discharge current can amount to several million amperes, from capacitors that can at present be charged to potentials up to 10 kV. The coil is unique for each part, while the size of the power source depends on the energy required for each joint and the number of joints per unit of time. The electrical conductivity of the material, the size of the workpiece and the overlap of the joint are all critical in determining the energy required. Copper and aluminium require less energy than steel. It is possible to weld different materials, e.g. copper to aluminium, aluminium to steel etc. The parts do not need to be circular: other shapes can be joined. In addition to welding, the method can be used for:

- Forming
- Calibration
- Crimping
- Punching

The method has the following benefits:

- It is a cold process
- It is contactless
- It has high precision
- It is extremely fast
- It is cost-efficient

The magnetic field outside the coil is not regarded as dangerous, as exposure is very short and the field strength declines with the square of the distance; in other words, very rapidly with increasing distance.

The process is best suited for volume production, as the coils need to be custom-made to suit the parts to be welded, which adds cost. The process is being evaluated by industries such as the automotive, white goods industries and electronics industry.

11.8 Cold pressure welding

Most welding methods involve the use either of heat alone (fusion welding) or a combination of heat and pressure. Cold pressure welding is an example of a welding method that is carried out entirely without heating. The principle involves pressing the parts of the workpiece together with such force that plastic deformation causes any residual oxide layers to be pressed out and a metallic bond to be made. The method is very suitable for use with copper and aluminium, and is often used for electrical connections. An example is connection terminals made of copper which can be cold pressure-welded to aluminium conductors. This is an application for which fusion welding is definitely not suitable, as it forms a brittle intermetallic compound with unacceptably low strength.

Grease and oxides must be removed from the surfaces immediately before welding, e.g. by wire brushing. The thin layer of grease applied by touching with the fingers is sufficient to prevent the necessarily intimate contact between the workpieces. However, aluminium can be successfully welded with its oxide layer on the surface since, because it is brittle, the substantial plastic deformation breaks it apart.

Welds can be made as butt welds or as overlap joints. Wire and rod materials are welded in the size of 0.5–12 mm. A toothed tool is used for overlap joints.

11.9 Diffusion welding

Diffusion welding is a method of joining surfaces to each other without melting and without deformation. The process is carried out under vacuum or in a protective gas atmosphere, with the application of high pressure and temperature over a relatively long period of time. Provided that the surfaces are clean, flat and accurately machined, large areas can be bonded in this way. A disc of tool steel, for example, with integral cooling passages, can be produced by welding a cover disc to another disc in which the necessary channels have been machined.

Diffusion welding can be used for many materials as titanium, nickel and aluminium alloys, including the joining of different types of metals and also the bonding of metals to non-metals. However, results are often improved by incorporating an intermediate layer between the two outer different layers.

11.10 References and further reading

N. Ahmed, *New developments in advanced welding*, Woodhead Publishing Limited, 2005.

D. Lohwasser and Z. Chen, *Friction stir welding*, Woodhead Publishing Limited, 2009

12 Other methods of welding

12.1 Introduction

This chapter reviews a range of welding techniques:

- electroslag welding
- electrogas welding
- stud welding
- laser welding
- electron beam welding
- thermite welding

Laser welding and electron beam welding can be categorised as power beam processes. Electrogas welding is a development of electroslag welding. Stud welding refers to a series of methods for joining materials using studs or bolts. Thermite welding is unique amongst welding methods in relying on a chemical reaction to achieve a weld.

12.2 Electroslag welding

Electroslag welding is a mechanised method for making vertical and near-vertical welds, with a maximum slope of 15° from the vertical. It is intended for welding very thick materials (from 40 mm upwards), although it may also be used for thinner materials.

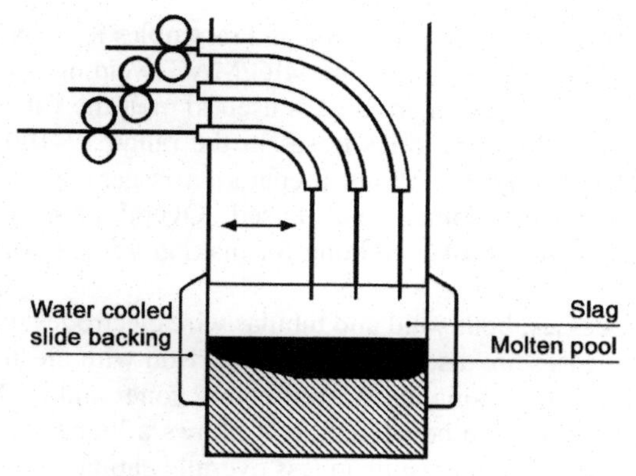

Figure 12.1 Electroslag welding.

The process is started by striking an arc between the electrode and the workpiece (see Figure 12.1). Flux is fed into the joint, and melts to form a bath of slag that increases in depth as more flux is added. When the temperature of the slag, and therefore its electrical conductivity, has increased sufficiently, the arc is short-circuited and the current is carried by the molten slag, maintaining its temperature by resistive heating.

The molten metal is prevented from escaping from the joint by water-cooled copper shoes, which may be fixed or arranged to move with the welding head. The weld is formed between these and the surfaces of the joint. The welding head moves up the joint as welding progresses. One or more filler wires may be used, depending on the thickness of the plate. If the material is very thick, the welding head may weave.

A variant of electroslag welding is one using a consumable guide. The filler wire in this version of electroslag welding is guided into the weld pool through a tube that melts and contributes to the filler material as welding proceeds. The tube may be coated in order to provide sufficient slag to keep the depth of the slag pool constant. The advantage of this method is that the welding head can be fixed and the length (height) of the weld determined by the length of the tube, which can be up to a metre long.

The benefits of electroslag welding include:

- High productivity
- Low cost for joint preparation
- Single pass, regardless of the plate thickness
- No angular deformation when making butt joints
- Little transverse shrinkage
- Little risk of hydrogen embrittlement.

A drawback of the method is the high heat input which means that the weld and HAZ cool slowly, thus allowing time for substantial grain growth. The basic material has insufficient toughness in the HAZ if there are requirements on impact strength of weld structure.

12.3 Electrogas welding

Electrogas welding is a development of electroslag welding and resembles it, in terms of arrangement and use. However, a shielding gas (as in MIG/MAG welding) is used instead of a pool of slag to protect the weld, and arc heat is used to melt the filler wire rather than resistance heat. It is used for plate thicknesses in the range 12–100 mm, using a weaving motion for greater thicknesses. Joint preparation is generally in the form of a simple I-joint and gap, although V-joints are also used. Overall process costs are considerably cheaper than manual MIG/MAG welding for making vertical joints in items such as large tanks.

As with other gas metal arc processes, both solid and tubular wire electrodes may be used. The same types of shielding gases are also used. In comparison with electroslag welding, electrogas welding produces a smaller heat-affected zone and a better toughness. A long filler wire stickout can be beneficial, as it allows a higher welding speed, melts less of the workpiece material and results in less overall heating.

12.4 Stud welding

Stud welding is the common name applied to several different methods of securing bolts, studs etc. to a workpiece. Compared with processes such as drilling and tapping, it is fast and simple. The commonest method is to heat the parts to be joined briefly but intensively by means of an arc. This melts the surfaces to be joined. The bolt or stud is then pressed into position with a special welding gun. Steel bolts or studs up to about 25 mm diameter can be welded in this way. The process may also be used for stainless steel, copper and brass items. Aluminium can be welded using the capacitor discharge method. Stud welding can be easily undertaken as a manual process and can also be easily mechanised, e.g. for robot welding.

In addition to the stud welding gun, the process requires a suitable power unit and control equipment for striking the arc and controlling the various operations. The power unit can be of a conventional type, or may consist of a group of capacitors which, after charging, provide a very fast discharge.

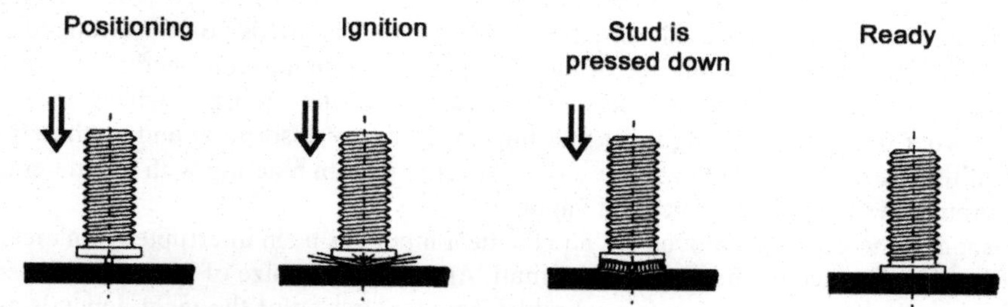

Figure 12.2 Stud welding by the capacitor method.

Capacitor Discharge (CD) stud welding (Figure 12.2) welds studs with a small stud on the head, which burns away and ignites the arc. Arc duration is very short, of the order of only a few milliseconds. This method is very suitable for welding on to thin metal, as there is little dispersal of heat to the surrounding metal, which means that the surface finish or treatment on the reverse side is not destroyed. The welding time is 2–5 ms, and studs of up to 8 mm diameter can be welded.

Conventional power units are suitable for bigger studs (see Figure 12.3). This method involves striking the arc by touching the stud against the workpiece and then lifting it. When the surfaces have been sufficiently melted, the stud is pressed against the metal. A ceramic ring is positioned around the stud, serving the dual purpose of containing the melt and protecting the process. Shielding gas can also be used. The minimum plate thickness is a quarter of the stud diameter, and the welding time is about 0.1–2.0 seconds. The surface of the metal does not need to be clean.

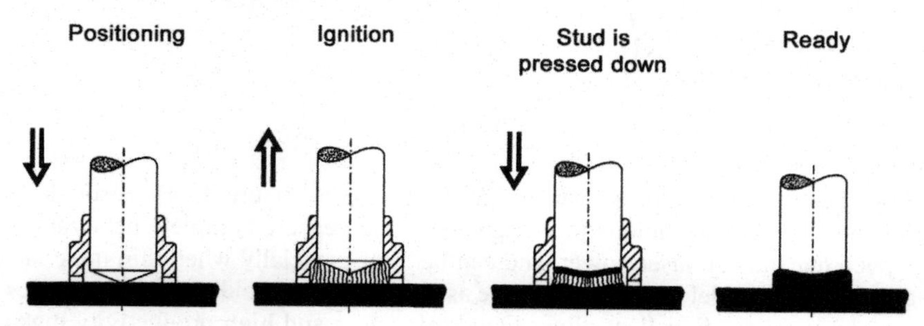

Figure 12.3 The arc method.

A variant of the arc method is the short cycle method. The welding time is shorter, generally less than 0.1 second, and the method can be used both with and without a ceramic ring and/or shielding gas. The sheet can be thinner in proportion to the diameter of the stud (down to 1:8), and the process is not sensitive to the surface coating of the sheet, e.g. galvanizing material, grease or mill scale.

12.5 Laser welding

Laser light possesses several unique properties, among which is the fact that it is parallel and highly concentrated. It can therefore be conducted, by mirrors or glass fibres, to a welding position which is remote from the power unit. It is also monochromatic, i.e. has a single definite wavelength, which depends on the type of laser used.

The laser beam is focused by a lens or mirrors into a point only a few tenths of one mm in diameter in order to provide a high energy density. The focus point is arranged to fall on, or slightly below, the surface of the workpiece. The material immediately melts and, some is vaporised. The vaporised metal in the hole forms a plasma which, being a good absorber of the incident light, further improves energy absorption and so the efficiency of the process. Shielding gas is used to prevent air from reacting with the material and to protect the lens from spatter and vapour.

As soon as the beam has moved on, and the heat input has been interrupted, the metal solidifies quickly since the heated zone is small. As a result, the size of the heat-affected zone is also small, and distortion is negligible. The penetration of the weld depends on the laser output power. Filler materials can be used, most often in laser-hybrid welding (see below).

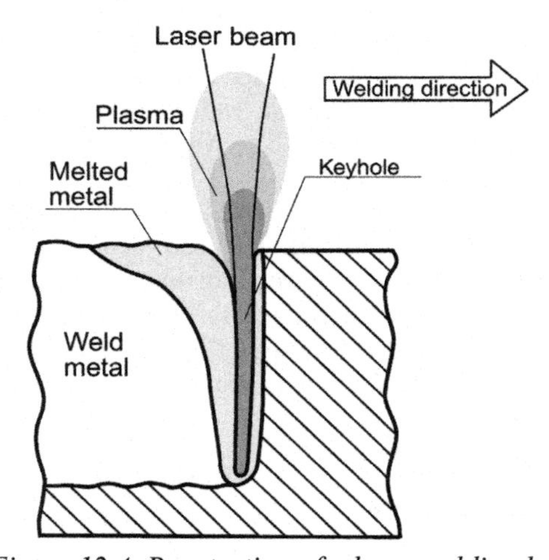

Figure 12.4 Penetration of a laser welding beam.

The laser beam produces deep, narrow penetration (see Figure 12.4), particularly with the keyhole welding technique (see Page 73), which ensure complete penetration. Welding speed depends on laser power, and can be high especially when thin materials are being welded. Laser welding is about twice as fast as plasma welding and eight times faster than TIG welding. Excellent dimensional tolerances and high productivity make the method superior to most others in many applications. In addition, laser welding is clean and quiet.

Laser welding is often used for welding materials that can only tolerate a low heat input, e.g. certain stainless steels and hardened materials, or for welding relatively thin sheets. The method is also used where complex parts require high precision. One of the main application areas for laser welding is the automotive industry. The technique is also used in aerospace and electronics industries and in general engineering. More powerful lasers increase the degree of penetration and trials of laser welding are being carried out in shipyards.

A drawback of laser welding is that low width/depth ratio of the weld geometry. In certain cases this may result in thermal cracking. In addition, hardenable steels can be locally hardened by the rapid rate of cooling. As the laser beam is only a few tenths of a millimetre in diameter, the method is tolerance-sensitive and therefore requires highly accurate jigs and fixtures. However, thinner materials can often be overlap-welded, which reduces the level of accuracy required. Another drawback of laser welding is the risk of undercutting which seriously reduces fatigue strength. Investment costs are high, but prices are coming down and laser welding can be expected to be more widely used in the future.

The resistance of a material's surface to wear and corrosion may be improved by applying a suitable cladding layer, either in small localised areas or over the entire surface. Laser techniques are well suited to cladding. Laser cladding can be used both in the manufacture of new components and for repairs, wherever an appropriate surface cladding is required. It produces a thin layer of pore-free weld metal, with a good surface finish and little dilution with the base material. Less energy is required than for other cladding methods, which means that there is less thermal distortion and a more finely grained structure. The cladding material is usually in the form of a flux bed but other forms are possible including wire, foil, chips etc. Welding is carried out under a shielding gas to prevent the formation of oxides.

The commonest types of welding lasers are the CO_2 laser and the Nd:YAG laser, the latter tending to be used for thinner materials and the former for thicker. The laser beam may be either pulsed or continuous. Other types of laser welding are fibre and diode lasers.

The CO_2 laser

The CO_2 laser generates its light in a tube through which a mixture of gases (including CO_2) flows, producing a wavelength of 10.6 μm. The energy input is by means of an electric discharge through the gas. It can produce a high power output, and is therefore popular for welding and cutting applications. The light is usually conveyed to the welding head and focused by mirrors. A shielding gas (often helium or an argon/helium mixture) is used to protect the lens and the weld. This helps to limit the amount of energy-absorbing plasma formed above the surface of the joint. Helium is preferred for this purpose, due to its high ionisation energy.

The CO_2 lasers available today are more powerful than Nd:YAG lasers. As welding speed is proportional to output power, CO_2 lasers can weld more quickly. The higher power also means that CO_2 lasers can weld plate up to 26 mm thick. However, a drawback is the high proportion of the beam energy that is reflected by certain materials. Different wavelengths are absorbed to different extents, which means that the light from CO_2 lasers is reflected more readily than that from YAG lasers. This is particularly noticeable when welding materials such as aluminium or magnesium alloys. Gold, silver and copper are also difficult to weld with CO_2 lasers.

The total running cost is comprised of the costs of electric energy, shielding gas, the laser gas and the maintenance of the optical system e.g. the mirrors. The total efficiency is low, only around 10 %. For personal protection, a pair of simple glasses is sufficient to absorb light from a CO_2 laser and to protect the cornea from being damaged.

The Nd:YAG laser

The active substance in this laser is neodymium (Nd), in the form of a dopant in a transparent rod of yttrium aluminium garnet (YAG). Energy is supplied by a flash tube or laser diodes. The light output wavelength is 1.06 μm, i.e. considerably shorter than that of the CO_2 laser, but still within the invisible infra-red section of the spectrum. An important difference is that the shorter wavelength enables the light to be carried by fibre optics and focused through ordinary lenses. This gives substantial practical benefits and makes it possible to use the laser for robot welding. Problems due to the presence of absorbing plasma are reduced when welding with Nd:YAG lasers, which means that argon and argon/CO_2 gas mixtures can be used. Acceptable results can even be obtained without a shielding gas when making spot welds or welding at low power.

This type of laser is particularly suitable for welding materials which would otherwise be difficult to weld, such as tantalum, titanium, zirconium, Inconel etc. One drawback is that YAG lasers are not available with such high power outputs as the CO_2 laser, which means that its use tends to be limited to metal thicknesses up to 6 mm. However, advances in increasing output and its combination with fibre optic light conductors makes this type of laser potentially very attractive.

It is important to note that the light from a Nd:YAG laser requires special protective goggles, as even the reflected light from the workpiece can damage both the cornea and the retina, as well as cause cataracts in the vitreous body of the eye. This is because the lens of the eye focuses the light on the retina. Even with protective goggles, it is not recommended to look other than fleetingly at anything in the vicinity of a Nd:YAG laser beam while welding is in progress.

Diode lasers

High-power lasers are bulky, and often have very poor efficiency. Laser diodes offer an interesting alternative. These are, in principle, the same as those fitted in every CD player. By grouping a large number of diodes, it is possible to deliver an output power of several kW. The wavelength is short, at about 800–900 nm, which gives good absorption in comparison with, for example, CO_2 lasers, and also allows the light to carried by optical fibres. However, the beam quality is still not good, and it is difficult to concentrate the light sufficiently, which reduces performance for keyhole welding or the cutting of thicker materials. Nevertheless, diode lasers are very suitable for heat conduction welding (without keyhole), cladding, soldering and welding of plastics.

Fibre lasers

A fibre laser uses a doped glass fibre as the laser medium. The energy source can be a diode laser, from which the light is reflected into the doped fibre by an outer sheath on the fibre. The advantage of this arrangement is the good beam quality. The essential combination of high power and a concentrated beam can be achieved with a higher efficiency than that of CO_2 or Nd:YAG lasers. The wavelength, from a fibre doped with ytterbium, is 1.07 μm.

Laser-hybrid welding

Hybrid welding refers to a combination of two welding methods, usually laser welding and an arc welding method such as MIG or plasma welding (see Figure 12.5). In combining a laser with MIG/MAG welding, the wire provides molten material that fills the joint, so reducing the requirements for exact positioning and accurate joint tolerances that would otherwise be required for laser welding alone. When welding fillet joints, this combination also provides filler metal to the joint. In comparison with ordinary MIG/MAG welding, process stability is improved, penetration is deeper and welding speed is considerably higher. Combining the two techniques also reduces the risk of undercutting, a drawback with laser welding which seriously reduces fatigue strength.

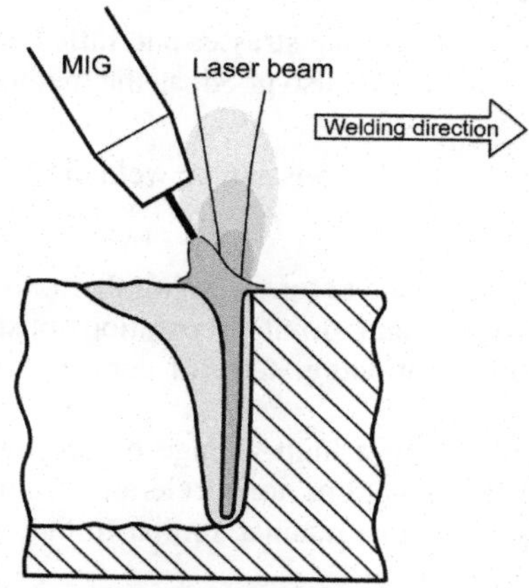

Figure 12.5 Principle of laser/MIG hybrid welding.

The method is suitable for making both butt and fillet joints and is mainly used for welding plates of 10 mm and thicker. The improved penetration means that fewer passes need to be made and less joint preparation is required than with pure (autogenic) MIG/MAG welding. The reduced heat input also produces lower residual stress and distortion.

12.6 Electron beam welding

Electron beam welding uses a very high-energy electron beam to produce deep, narrow penetration. The electron beam has a higher energy content than a laser beam, and is also smaller. Penetration is deeper for a given level of power, and the overall efficiency of the energy conversion process from input electricity to output beam power is much higher. Important characteristics include the high energy density, which makes it possible to melt the gap between two parts without the problem of distortion.

Welding normally has to be performed in a vacuum as the electron beam is absorbed by air. This complicates the process when changing the workpiece. On the other hand, the absence of air is good for the welding process, as there are no reactions between the air and the metal of the weld or the workpiece. Welds are normally made as butt welds.

Electron beam welding is often used for advanced materials and complex, critical parts such as turbine rotors, but it can also be suitable and economic for many simpler processes involving large production runs. It is very suitable for butt welding materials of different thicknesses, but is particularly competitive for welding thick materials of up to 250 mm. Electron beam welding has the following advantages:

- The high flexibility of this method permits everything from thin sheet to very thick materials to be welded.
- Welding speed is much higher than (for example) arc welding. However, when assessing overall productivity, allowance must be made for the time required to evacuate the air from the vacuum chamber.
- The vacuum means that reactive materials such as titanium can be welded without risk of oxidation.
- Welds are narrow, with a high penetration/depth ratio.
- Heat input is low in absolute terms, resulting in low residual stresses and little distortion of the workpiece. Reproducibility and tolerances are also good, as the method is mechanised.
- Many otherwise difficult materials and material combinations can be welded.
- The method is perfect for sealing vacuum chambers.

It may be necessary to demagnetise magnetic materials before welding them, as the magnetic field could otherwise deflect the electron beam. Joint preparation must be carried out carefully, including ensuring accurate positioning. Careful control of the beam track along the joint is also essential.

The electron gun (see Figure 12.6), is supplied from a high-voltage power source (30–175 kV), but at a low current (less than 1 A). The electrons are accelerated from the anode and are focused and deflected by magnetic coils in a manner similar to that used in television and computer screens.

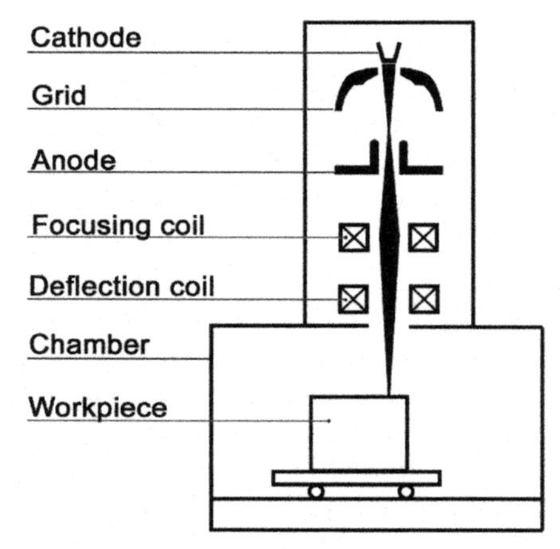

Figure 12.6 Electron beam welding.

The electron beam requires a vacuum, and so welding is carried out in a vacuum chamber. This normally requires the chamber to be opened to load or change a workpiece, after which the air must be evacuated by a high-vacuum pump. However, various designs that use forms of air locks for the loading and unloading of materials have been developed to deal with smaller items or (for example) strip materials. When the electron beam meets the workpiece, it produces a secondary emission of x-rays, and so the vacuum chamber also provides protection against this radiation.

Although the electron gun itself requires a high vacuum, the vacuum in the rest of the chamber does not need to be quite so high. The electron beam can even travel a short distance in air, but is quickly absorbed and scattered, so limiting penetration. In addition, if the welding is performed outside the vacuum chamber, some other method of protection against x-rays will be required.

Welding is normally performed by traversing or rotating the workpiece by programmable control, with the electron beam remaining stationary. Particular attention must be paid to maintaining appropriate accuracy, bearing in mind the narrow beam and joint. However, this can be achieved by deflecting the beam to make it sweep back and forth across the joint (though this affects the penetration profile) or by joint tracking.

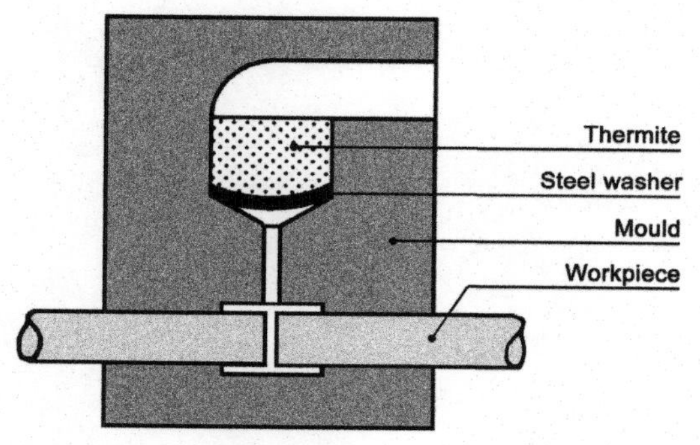

Figure 12.7 A mould set up for thermite welding.

12.7 Thermite welding

The energy source for this method is the heat released by the chemical reduction of a metallic oxide. This means that welding is performed by a chemical reaction, without any input of external energy.

The parts to be welded are positioned in a specifically shaped mould as shown in Figure 12.7. The welding powder, which is known as thermite, and consists of a metal oxide, such as iron oxide mixed with aluminium powder, is loaded into a compartment above the workpiece. As aluminium has a much higher affinity for oxygen than iron, an exothermic reaction occurs:

Iron oxide + aluminium => Iron + aluminium oxide + heat

The powder is ignited by a spark, resulting in the oxide being reduced, a large amount of heat being released and the iron melting. The very high-temperature melt, consisting of molten iron at about 2 500 °C, runs down into the mould, melts the joint surfaces and forms a welded joint with the workpiece.

The process is commonly used for joining rails and uses a thermite mix consisting of iron oxide and aluminium. Another application is the joining and connection of copper conductors, which uses a powder consisting of cuprous oxide, aluminium and tin. This mixture can also be used to join copper conductors to steel. A powder for welding aluminium is also available.

12.8 References and further reading

J. Lawrence et al. (eds), *Advance in laser materials processing technology*, Woodhead Publishing Limited, 2010.

F. Olsen, *Hybrid laser-arc welding*, Woodhead Publishing Limited, 2009.

A. Ahmed, *New developments in advanced welding*, Woodhead Publishing Limited, 2005.

TWI, *Laser welding*, Woodhead Publishing, 2000.

13 Cutting methods

13.1 Introduction

In the literature, thermal cutting processes by gas, plasma or laser are often covered in conjunction with the corresponding welding methods. This is because almost the same equipment is used for both processes, and also because the methods are often utilised together. It is also appropriate to cover competing methods such as water jet cutting.

This chapter also describes air carbon arc gouging and oxy-fuel gas flame gouging; methods which are often closely associated with the corresponding welding processes.

13.2 Thermal cutting

Thermal cutting is widely used in connection with the preparation of parts for welding. In addition to cutting plates etc., it may also be necessary to prepare the joints by bevel-edging them. The quality and smoothness of the resulting cut surface are generally satisfactory for the purpose, and the methods are easy to mechanise.

Oxy-fuel cutting

Oxy-fuel cutting uses a flammable gas, generally acetylene or propane. Burning the gas in oxygen, rather than just in air, produces a flame with a high temperature. The flame first preheats the workpiece: when a sufficiently high temperature has been reached, a jet of oxygen produces the cut by actually burning the metal. This produces a metal oxide in the form of liquid slag, which is blown out of the joint by the jet of gas.

The flame also helps to keep the upper surface of the plate above the ignition temperature of the metal while cutting is in progress, although most of the necessary heat required for the cutting comes from combustion of the actual material being cut. For example, when cutting 25 mm steel, about 85 % of the heat comes from combustion of the iron. In thinner materials, however, a greater proportion of the heat is applied by the flame.

For successful gas cutting, the material to be cut must fulfil certain conditions:

- The oxide must have a melting point that is lower than the melting point of the metal itself. In the case of iron, the oxides melt at about 1400 °C, which is lower than the 1530 °C melting point of low-carbon iron. It is the melting temperature of the oxides that explains why stainless steel and aluminium are not suitable for gas cutting.
- The ignition temperature of the metal must be lower than its melting point. In the case of low-carbon steel, the ignition temperature is about 1050 °C.
- Combustion of the metal must create sufficient heat to maintain combustion.

From this, it follows that it is only low-alloy steels with a carbon content of up to about 0.3 %, that can be cut in the usual way by oxygen burners. In such steel, thicknesses up to about 300 mm can be cut by oxy-fuel cutting. Where quality requirements in respect of the finished cut are less stringent, very considerable thicknesses can be cut, up to about 3000 mm. At the other end of the scale, below 25 mm, oxy-fuel cutting is in competition with plasma cutting, which gives a higher rate of cutting, as shown in Figure 13.1.

Other methods of cutting are recommended for cutting stainless steel and cast iron. However, oxy-fuel cutting can be used to cut these materials, with the use of appropriate additives.

Oxy-fuel cutting can be carried out either manually, using a cutting torch, or by machine, with a numerically-controlled cutter head. As in gas welding, there are two main types of torch: high-pressure torches and injector torches. The difference between welding torches and cutting torches is that the latter have a nozzle for the oxygen cutting jet, generally in the centre of the flame nozzle.

The typical cutting speeds for plasma cutting, oxy-fuel gas cutting and laser cutting are shown in Figure 13.1.

Gases in oxy-fuel cutting

As described in the chapter on gas welding, *acetylene* has special characteristics, burning with an intense flame and a high combustion velocity. This high combustion velocity can be a disadvantage or even a danger, in that the combustion front can migrate backwards into the burner nozzle. On the other hand, the hot core of the flame makes the gas suitable for cutting thinner materials, which can be done with good productivity.

Propane burns with a flame with lower heat concentration. This therefore spreads the heat more evenly along the cut, which can be an advantage when cutting thicker materials.

Hydrogen is not commonly used as a fuel gas, although interest in it has been aroused through the ability to produce oxygen and hydrogen by hydrolysis of water. This involves the use of electrical energy to dissociate water into its constituent elements, at the rate required for cutting. In terms of their thermal characteristics, hydrogen flames are similar to propane flames.

Oxygen performs three duties in connection with gas cutting. It produces the heating flame with the fuel gas; it burns (oxidises) the material to be cut and it blows the resulting slag out of the cut. The purity of the oxygen is very important for cutting speed: 99.3–99.7 % purity is common. A reduction of 0.5 percentage points reduces the cutting speed by about 10 %.

Plasma cutting

The hot concentrated jet produced by the plasma method, which has previously been described in connection with welding, is very suitable for cutting. However, as opposed to gas cutting, which works primarily by burning the material using the oxygen in the cutting jet, plasma cutting works by melting the material and then blowing the molten material out of the cut by the pressure of the plasma jet. When used for cutting, the pressure of the plasma gas is higher than that used for welding. Both smoke and noise are generated, at least when cutting thicker plate. However, this can be considerably reduced by cutting the metal on a cutting table, with the metal itself under water. The range of applications is wide, although materials of particular interest for this process are those such as stainless steel, aluminium and copper, which cannot be cut by ordinary oxy-fuel gas cutting.

Plasma cutting can be the best alternative, even for ordinary low-alloy steel. For thicknesses up to about 25 mm, the rate of cutting can be considerably faster than that of oxy-fuel gas cutting, which also means that the size of the heat-affected zone is reduced. On the other hand, the kerf width is about 1.5–2 times wider than that produced by gas cutting. In addition, the surfaces are not completely parallel, the cut being slightly wider at the top.

New designs of plasma cutters have been developed in order to solve this problem. A method known as precision plasma (High-Tolerance Plasma Arc Cutting, HTPAC, in American literature) has been developed. Its features include a smaller gas nozzle opening and a narrower cutting jet, with a higher current density. Pure oxygen is used when cutting low-alloy steel. The resulting characteristics of precision plasma cutting comes closer to those of laser cutting. Greater accuracy of control of the cutting nozzle is required when cutting thin metal, in order to meet the tolerance requirements. This also includes careful height maintenance above the surface of the workpiece.

A hand torch can be used for simpler jobs, although industrial production generally uses numerically controlled cutting tables, with one or more cutting heads, approximately as for gas cutting. Noise, visual radiation/arc glare and smoke can be quite intensive, but can be considerably reduced if the metal to be cut and the plasma nozzle are under water. The arc electrode is normally tungsten or tungsten with thorium oxide. However, the development of electrodes containing hafnium or zirconium has made possible the use of oxidising cutting gases, even to the extent of using ordinary air.

The power unit has a constant current characteristic, as for TIG and plasma welding, but must be designed for a much higher voltage. The operating voltage exceeds 100 V, and the open-circuit voltage can exceed 200 V. Special measures must be taken to prevent the operator from coming into contact with these dangerous voltages.

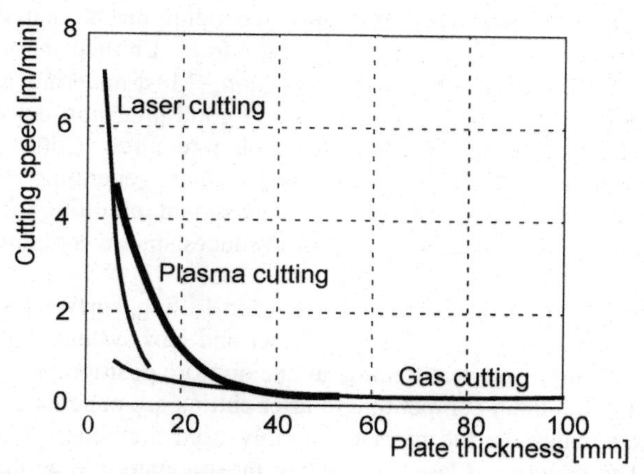

Figure 13.1 Typical cutting speeds for plasma cutting, oxy-fuel gas cutting and laser cutting.

Plasma gases

Pure argon is sometimes used as the pilot gas in order to ensure reliable ignition of the pilot arc. (The pilot arc is struck before the cutting operation starts, but is not transferred to the workpiece, i.e. it strikes between the central electrode and the plasma nozzle). The cutting gas must have good heat transfer properties: pure nitrogen, mixtures of argon/hydrogen, nitrogen/hydrogen or just compressed air are used.

One way of increasing the cutting speed in low-alloy steels is to use an oxidising gas that provides an active contribution by burning the metal, as with oxy-fuel gas cutting. The simplest and cheapest gas, of course, is ordinary compressed air. However, this imposes special requirements in respect of the electrode. As a tungsten electrode is attacked by oxygen, so a well-cooled hafnium or zirconium electrode must be used. It must also be accepted that there will be somewhat higher costs for replacing the electrodes.

Laser cutting

A laser beam has excellent characteristics for cutting: in particular, the precision of the cut is very good, and there is very little thermal effect. The method is best suited for relatively thin materials, where very high productivity is required. Many non-metallic materials can also be cut by laser.

Using a lens or a mirror, the laser beam is focused to a 0.05–0.25 mm diameter spot on the workpiece: the small size means that the energy density of the radiation on the spot is high. As in laser welding, the focal point of the beam is often positioned in the upper part of the workpiece. The high energy density of 10^7–10^{13} W/cm^2 in the beam instantly melts or evaporates the metal, which is blown out below the kerf by the cutting gas. When the laser beam or workpiece starts to move, the beam makes a kerf about 0.1 mm wide. As soon as the beam has passed the metal solidifies, as the actual heated volume is small. However, the molten zone must penetrate right through the workpiece for a successful cut.

Laser cutting produces straight, clean and narrow kerfs, with narrow tolerances and low material losses. Kerf width varies between 0.05-1 mm, depending on the material thickness. Cut surfaces are stress-free, and generally also burr-free. Limited thermal effect results in a small HAZ, and therefore little or no distortion. Most materials and geometries can be cut. The relatively high cutting speeds and good precision deliver high quality and productivity. Further heat treatment is very seldom required. Cutting is easy to control and automate, e.g. by CAD/CAM. The method is clean, generating relatively little noise, and requires no regular resharpening and replacement of cutting tools. The only material that should not be laser-cut is PVC, as it produces smoke containing toxic hydrogen chloride (HCl).

In addition to welding and cutting, lasers can also be used for drilling, surface hardening, heat treatment, marking and engraving. The same laser and lens system can be used, for example, for cutting, welding and heat treating at one or more position.

For thicknesses up to 10 mm, the main competitors to laser cutting are punching and nibbling machines. For thicker materials, the most commonly used methods are gas cutting and plasma cutting. The benefits of laser cutting are most apparent in applications where the use of punches or nibbling machines would be too expensive in relation to the numbers of parts to be produced, or where the material is too thin for conventional machining. Laser cutting is often used where high quality and/or tolerance requirements apply, or for difficult-to-cut materials such as ceramics. Applications in the automotive industry include the cutting of die-cast or moulded and reinforced polymer parts.

The laser light is generated by CO_2 or Nd:YAG lasers, and its characteristics are described in the section on laser welding. The laser itself is stationary, and the beam of light is carried to the cutting head where it is focused by a lens. The cutting motion may be two-dimensional or three-dimensional: the Nd:YAG laser is preferable for three-dimensional control, as it can provide higher output powers and the light can be conducted through glass fibres, with the cutting head controlled by an industrial robot.

Cutting with a CO_2 laser

The beam from CO_2 lasers is strongly reflected by certain metals, such as aluminium and particularly by polished copper, silver or gold. The cutting of aluminium is also complicated by its high thermal conductivity. The resulting cutting speed is low, and the maximum thickness that can be cut is less than that which can be cut by an Nd:YAG laser of the same power. CO_2 lasers can be used for cutting non-metallic materials such as glass wool, ceramics, wood, paper, artificial leather, concrete, nylon etc. Air is normally used as the cutting gas when cutting organic materials.

Cutting with an Nd:YAG laser

Laser cutting with an Nd:YAG laser produces excellent cutting quality, with sharp edges and a narrow kerf, so permitting high cutting speeds in thin materials. The Nd:YAG beam can be conducted by fibre optics, which makes the equipment more flexible, permitting cutting in various positions and at different distances from the laser source. The automotive industry, for example, uses Nd:YAG lasers for 3D cutting of vehicle bodies.

Other advantages of the Nd:YAG laser include very fine detail cutting of thin materials, and the ability to cut high-reflectivity materials such as copper or silver alloys. The method, often in combination with high-frequency pulsing, is also suitable for very precise cuts of relatively small sizes in difficult materials, such as diamond, ruby, sapphire, ceramics etc. Almost all metals and ceramics can be cut with Nd:YAG lasers, although there can be problems with non-metallic materials that are transparent at this wavelength, such as glass, quartz or many polymers.

The kerf surface quality is often much better when cut by an Nd:YAG laser, but cutting speeds are often lower than with a CO_2 laser, due to lower output power. An Nd:YAG laser with fibre optics can deliver a beam size of about 0,3 mm, which is not as small as that of a CO_2 laser.

In addition, Nd:YAG cutting is somewhat more expensive than CO_2 laser cutting. This is due partly to a higher equipment cost, and partly to a linear operational cost which is often somewhat higher. For cost and power reasons, therefore, more CO_2 lasers than Nd:YAG lasers are bought and used for industrial cutting.

Laser cutting gases

The cutting gas protects the lens from cut particles or splashes of molten materials, and blows away molten or evaporated material. The gas may be active or inert: oxygen or air are reactive, while argon and nitrogen are inert. For cutting most metals, an active gas increases the cutting energy input, but tends to leave oxidation products on the kerf face. The use of an inert gas produces a cleaner cut, but at the cost of cutting speed. Carbon steel and stainless steel can be cut with oxygen or nitrogen, while aluminium should be cut with nitrogen. Titanium should be cut with argon or argon/helium mixtures. Materials such as aluminium and plastic polymers can also be cut with ordinary air, as the oxygen in the air helps to burn off the sharp edges of the cut.

13.3 Water jet cutting

Water jets alone (i.e. without any additive materials in the water) can be used for cutting soft or porous materials. When cutting metals or hard materials, such as glass or stone, *abrasive water jet cutting* is used, with sand being an additive in the jet.

The unique feature of water jet cutting is that there is no thermal effect on the material, thus eliminating any thermal or mechanical stresses that could affect the results. This means that, although the cutting speed may be lower than that of certain competing methods, time can be saved overall through elimination of the need for any subsequent treatment. The quality of the cut is good when compared with the results produced by thermal methods. The method is also suitable for cutting materials that can be difficult to cut in any other way.

Most materials can be cut by water jets. Using water alone, materials such as wood, paper, felt, foamed plastic etc. can be cut. Abrasive water jet cutting can deal with metals such as stainless steel, copper, aluminium and titanium, or with composites such as glass fibre-reinforced plastics or with hard materials such as glass, ceramics and natural stone.

Water jet cutting is generally employed where other methods are unsuitable. The cutting speed when cutting steel less than 20 mm thick, for example, is only about one-tenth of the speed that can be achieved by plasma cutting. Metals can be cut at a rate of about 10–30 cm^2/minute, or somewhat more for soft metals. Thicknesses can be up to about 100 mm. Materials such as glass, plastic, rubber, stone etc. can be cut at rates of about 100–300 cm^2/minute.

An electrically driven hydraulic oil pump produces an oil pressure of 150–200 bar. The oil then drives a high-pressure pump which produces a water pressure of up to 4000 bar. This pressure is converted in the jet to a very high velocity of up to about 1000 m/s. The water jet nozzle is subjected to very high wear, and therefore includes an insert which is generally made from industrial sapphire in order to provide an operating life of perhaps about 100 h. The hole through the sapphire is usually about 0.1–0.3 mm, producing a hair-thin jet and a cut which is almost equally narrow. The cut produced by abrasive cutting is somewhat wider, up to about 1.5 mm.

A cutting table is generally used, with a numerically controlled cutting head, as in thermal cutting processes. Industrial robots may also be used for three-dimensional cutting.

13.4 Thermal gouging

Welding frequently requires the cutting away of surplus material. In such cases, gouging can be more efficient than grinding when repairing defects in welds or in cutting a groove to avoid weld defects when the workpiece is to be turned over and the weld completed from the root side.

Several of the methods, and particularly those using an electric arc, create considerable quantities of smoke, so that special ventilation should be provided when using them indoors. In addition, as gouging involves melting the material and blowing it away, the operator should protect himself and the surroundings.

Oxy-fuel gas flame gouging

This method is based on the same technology as that used for gas cutting. It uses a special nozzle, which facilitates working along the workpiece surface. It is suitable for use with carbon steel and low-alloy steels, in the same way as is gas cutting.

Air carbon arc gouging

This method has also gone under the name of *arc air gouging* or *carbon arc gouging*. It uses approximately the same equipment as welding with coated electrodes. The gouging electrode is copper-plated carbon, which is used in an electrode holder with an outlet for compressed air. The best power unit is one capable of providing a high current and if possible, also a high short-circuit current, in order to maintain the correct arc force. The electrode is held at an angle to the workpiece and, together with a jet of compressed air, can remove metal at a high rate.

The carbon electrode should be connected to the positive pole of the power unit: special electrodes are available for use with AC. Electrode diameters from 3 mm to 19 mm are used, depending on the current available and the desired rate of metal removal.

Manual metal arc gouging

This does not require any equipment other than that which is used for welding with coated electrodes. The electrode however, is a special gouging electrode, with a thick coating which produces a considerable quantity of gas. The arc is struck in the usual manner, and the electrode is then inclined at a considerable angle to the work, with the tip pointing in the direction of travel. It produces a smooth groove, with a high rate of metal removal. Best performance is obtained by connecting the electrode to the negative pole or by using AC.

Plasma arc gouging

Plasma arc gouging uses the same equipment and gases as plasma cutting. The nozzle, however, may incorporate a nozzle for shielding gas. It is held about 20 mm from the workpiece at an angle of approximately 45° and pointing in the direction of travel. The angle may be changed to vary the width/depth relationship of the groove.

Compared to the air carbon arc there are the following benefits:

- Decreased generation of fumes and gases and reduced noise levels
- Higher productivity and groove quality
- No risk of carbon pick-up
- Suitable also for non-ferrous metals

13.5 References and further reading

J.F. Ready. *LIA Handbook of laser materials processing*, Ch. 12, Laser cutting, 2001.

K. Nilsson. *Technical development of the laser process*, doctor's thesis, LTU, 2001:21.

14 Surface cladding and hardfacing methods

14.1 Introduction

The application of stainless material to a lower alloy steel (known as cladding) is an economic method of producing reactor pressure vessels or pressure vessels for such applications as the chemical industry, where a thick-walled vessel with internal corrosion protection is necessary. There are several different welding methods that can be used:

- Submerged arc welding using a solid wire electrode, often Twin Arc double wire, or using a flux cored wire.
- Submerged arc welding using a broad but thin strip electrode.
- Strip electrode welding with a flux that enables the process to be carried out using the electroslag principle, i.e. without an arc, but with the heat being generated by resistance heating in the molten slag. See "Electroslag welding" on page 133.
- Plasma cladding.
- Thermal spraying: see Page 153.

The aim is to achieve a sound weld, but with little melting of the underlying material. In this respect, the electroslag method is preferable to ordinary submerged arc welding, as it penetrates less deeply into the substrate material and so results in less mixing of the weld metal. Several of the ordinary submerged arc welding methods can be used, but it may be necessary to apply two or more layers until a sufficiently pure layer of welded metal is produced. Cladding is often performed by mechanised methods, as there are often larger areas to be covered.

If a hard surface is required with underlying material of a higher degree of toughness, the normal practice is to apply the hard surface layer by welding. This process is known as hardfacing. Application of high-abrasion-resistance alloys by welding is a method of repairing machine parts or other metallic items that are subject to abrasive wear. Examples include rails, turbine blades, excavator bucket teeth and conveyors in the mining industry. It is often cheaper to repair such parts than to replace them by new ones, as it is quicker and there is no need to hold stocks of spare parts. Hardfacing can be done either by automatic welding equipment or manually, depending on the circumstances and the amount of welding to be done.

14.2 Types of wear

The type of hardfacing material to be used depends entirely on the type of wear to be protected against. There are many different wear mechanisms, although they can be divided roughly into the following four classes for simplicity:

- Friction and adhesion
- Abrasion
- Impact and shocks
- Heat, oxidation and corrosion

Friction and adhesion occur when two metallic surfaces roll or rub against each other, e.g. a shaft and bearing or a cable and sheave. Even the most highly polished surfaces have a microscopic level of unevenness, which results in wear. Particles from both surfaces are pressed together and broken off, causing an increase in wear. The most important factors in this respect are the finish of the surface, contact pressure and material structure. In general, it can be said that wear particles from two surfaces having the same alloy and hardness 'combine' more easily resulting in greater wear. It is therefore desirable to avoid allowing two surfaces of the same material to come in contact.

Abrasion results when small, hard mineral particles rub across a metal surface and cut away particles from it. The harder the mineral and the sharper the edges of the particles, the greater the amount of wear. This type of wear mechanism occurs, for example, in dredging and in the transport of minerals. The material must be as hard as possible to counter abrasive wear, although its microstructure and surface are also important factors. Alloy steels containing chromium or tungsten are usually recommended.

Impact and shocks occurs in, for example, crushers and excavator buckets. If this is the main wear mechanism, it is important to use a steel that combines toughness and ductility, which is capable of absorbing the shock by deforming instead of by cracking. However, such wear often occurs in combination with abrasive wear, which requires a hard surface. Chromium and tungsten alloys, mentioned above, are relatively brittle, which, although unimportant when abrasive wear is the only mechanism, result in cracking if the material is exposed to impact. For applications such as crushers and hammers therefore, an alloy containing 14 % manganese is widely used, as it produces a hard surface with a ductile interior.

Heat, oxidation and corrosion may occur in tools used for hot working processes and for casting. These are exposed to cyclic thermal loads which eventually result in fatigue failures. Working in an oxidising environment produces a layer of oxide on the surface, which can then crack due to thermal expansion, thus exposing new metal for oxidation and allowing the process to continue until the part is entirely worn away. This particular form of wear is countered by the use of nickel and cobalt alloys, which have high resistance to abrasive wear, corrosion and thermal fatigue.

TABLE 14.1 Cladding materials.

Type of steel	Properties	Application / resists
Low alloy, low carbon	Tough	Building up / friction resistance
14 % manganese	Tough, work hardens	Shocks and impact
Martensitic	Tough and hard	Shocks and abrasion
Chromium carbide, tungsten carbide	Hard, brittle	Abrasion
Cobalt and nickel alloys	Hard at high temperatures	Tool steel / Corrosion, high temperatures

Appropriate cladding materials and welding methods have been developed for each of the four main classes of wear described above (see Table 14.1). However, a common feature is that hard alloys should not be applied in more than two or three layers, as their poor coefficients of thermal conductivity can result in the cladding cracking or separating from the underlying material. A further common requirement is that there should be as little mixing with the parent material as possible, in order to avoid degradation in the properties of the cladding. For this reason, a foundation layer of less hard alloys is often applied, with the fully hard surface layer applied on top of it.

14.3 Thermal spraying

Thermal spraying is used to apply metallic or ceramic layers to metals, for such purposes as producing a corrosion-resistant or wear-resistant layer on low-alloy steel, for making good material lost by wear and tear or machining error and for the application of electrical or thermal insulation. Layer thicknesses vary from about 10 μm up to a few mm, sometimes even tens of mm, depending on the application.

The method is particularly important for various types of repairs. Large and expensive shafts, blocks or rolls etc., which have been worn down outside their permitted tolerances, can be restored to a usable condition by a modest work input. Thermal spraying is also very suitable for use as a method of construction, where there is a need to apply various types of corrosion-resistant or wear-resistant layers to metal surfaces.

Thermal spraying requires an appropriate cladding material, in the form of wire or powder to be heated to its melting point by the thermal energy in a flame or arc, with the molten particles then being carried by a jet of gas on to the workpiece. The particle sizes are of the order of 50–200 μm.

Cladding layers applied by flame or arc contain about 5–15 % by volume of pores, and up to 5–15 % of oxides. The porosity and oxide contents of layers applied by high-velocity flame spraying and (in particular) plasma spraying are lower. The strength of the cladding depends on its material analysis, pretreatment of the workpiece and the spraying method used. When spraying steel, the strength is low in the spray direction (20–80 N/mm^2), but higher in the longitudinal direction (80–150 N/mm^2).

Spraying is carried out using a flame spraying gun of a size that can be handled manually. Nevertheless, it is often mounted on a support, with the workpiece arranged to rotate or travel in front of it. This is essential when thin, uniform layers are to be applied.

Pretreatment of the workpiece can involve degreasing and roughening of the surface by blasting, rough turning or grinding. The aim is to achieve a matt surface with an appropriate key. Turning creates a corrugated surface, thus increasing the surface area and improving the adhesion of correctly applied cladding. A bonding layer of special material is often applied before the final top layer. The bonding layer should be thin and even, but with good coverage.

Arc spraying, high-velocity flame spraying and plasma spraying all produce very high noise levels, which means that operators must wear hearing protection, and that the use of acoustically insulated working areas is recommended. The use of a spray booth with extraction ventilation to deal with smoke, gases and surplus powder is also advisable.

The most important methods of thermal spraying are:

- *flame spraying,*
- *high-velocity flame spraying,*

- *arc spraying,*
- *plasma spraying,*
- detonation spraying.

Each has a somewhat different application area, depending on the type of material, cost and performance. All can be used for spraying metallic materials, but not all are suitable for spraying non-metallic materials.

Flame spraying

The heat source for flame spraying is a flame, produced (as for gas welding) by combustion of acetylene or propane in oxygen. The cladding material, in wire or powder form, is fed continuously into the flame, where it melts, and the molten particles are then blown on to the workpiece by a jet of compressed air as shown in Figure 14.1.

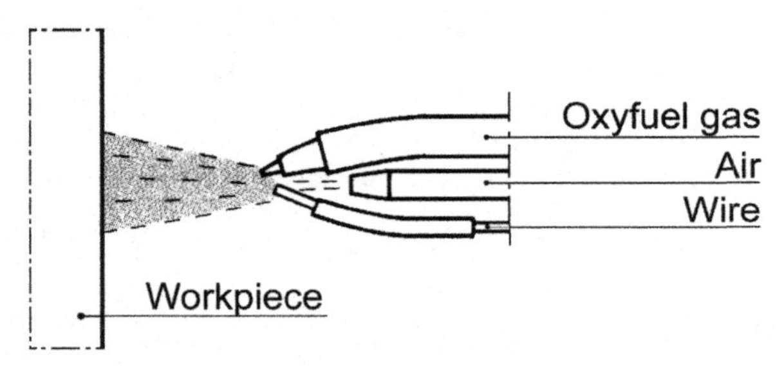

Figure 14.1 Schematic diagram of flame spraying with wire.

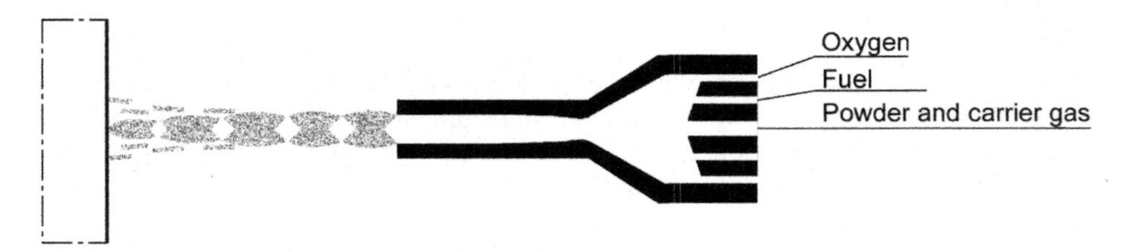

Figure 14.2 Schematic diagram of high-velocity flame spraying.

High-velocity flame spraying

High velocity flame spraying is often also referred to as HVOF spraying (high velocity oxy-fuel spraying). The method, which has developed rapidly in recent years, produces a high velocity of the molten particles, giving a very strong, dense layer, with less than 1 % porosity and low oxide inclusion.

Fuel gases may be paraffin, LPG or hydrogen, burnt with oxygen at high pressure in a combustion chamber. Temperatures of 2700–3100 °C are reached, depending on the fuel gas. The gas leaving the combustion chamber is accelerated to supersonic velocity (1500–2000 m/s) in an appropriately shaped nozzle. The high velocity produces characteristic shock waves in the flame, which are visible as a diamond pattern (see Figure 14.2). The gun is often water-cooled. The cladding material is generally applied in powder form which is fed into the nozzle by an inert gas as shown in the schematic diagram in Figure 14.2.

Arc spraying

This method uses an electric arc as the heat source which is struck between the tips of two sacrificial electrodes made of the cladding material. The electrodes are fed continuously into the arc, where they melt, and from where the molten particles are blown onto the workpiece by a jet of compressed air or gas as shown in Figure 14.3. The cladding material is always a metal, as it must be electrically conductive. Different materials can be used in the two electrodes, to produce a cladding which is a mixture of both.

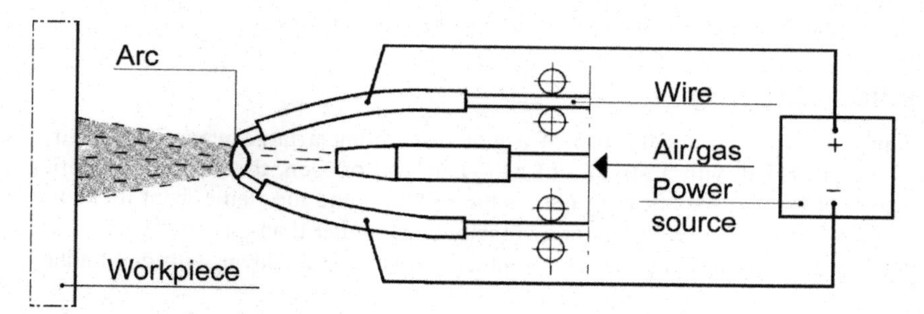

Figure 14.3 Schematic diagram of arc spraying.

Arc spraying generally produces the best adhesion to the workpiece and also offers the lowest production costs, apart from the spraying of zinc-rich materials. However, some alloying constituents, particularly carbon may be vaporised, which complicates the spraying of high-carbon cladding materials.

Plasma spraying

Plasma spraying often uses an arc as the heat source, which is struck between two non-sacrificial electrodes. A flow of gas past the arc blows the arc plasma out through a nozzle. At 10 000–20 000 °C, the plasma temperatures are higher than the temperatures used in other spray methods. Gases commonly used are H_2, N_2, Ar or He.

The cladding material is supplied in the form of wire, bare or powder, and fed into the arc, where it melts and is conveyed to the workpiece by the plasma jet as illustrated in Figure 14.4. Powder is the most commonly used material. The high temperature of the plasma enables ceramics and metal oxides with high melting temperatures to be sprayed. The equipment is more expensive to buy and run than that used in the methods described above, so plasma spraying is less often used for spraying simpler materials with a lower oxidation sensitivity. Its main application is the spraying of non-metallic materials as cladding for metallic materials with electrically insulating layers.

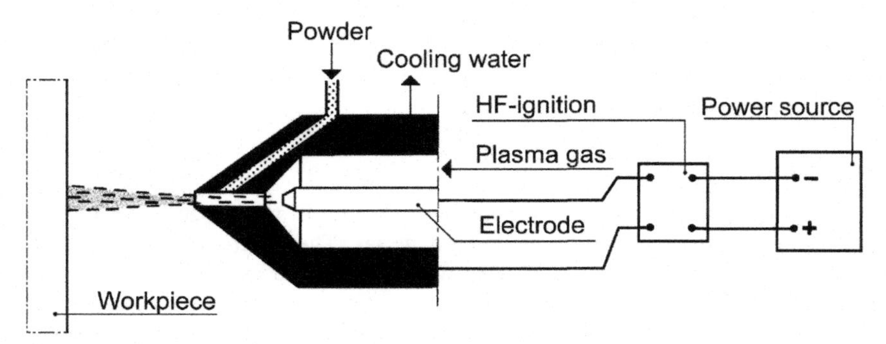

Figure 14.4 Schematic diagram of plasma spraying.

For optimum process control, plasma spraying can also be carried out in a vacuum chamber or a chamber with a suitably controlled atmosphere.

Detonation spraying

Detonation spraying involves the use of a gun resembling a large water-cooled rifle barrel, which is loaded with a mixture of acetylene and oxygen, the cladding material being in powder form. A spark plug ignites the mixture, and the flame front melts the powder driving it onto the workpiece at a velocity even higher than that of HVOF. This is repeated at a cycle rate of about 5 Hz, producing a very dense layer with good adhesion.

14.4 References and further reading

H. Dong, *Surface engineering of light alloys*, Woodhead Publishing Limited, 2010

B. Mellor, *Surface coatings for protection against wear*, Woodhead Publishing Limited, 2006

TWI, *Surfacing*, Woodhead Publishing Limited, 2000

15 Mechanisation and robot welding

15.1 Introduction

Mechanisation of arc welding can be applied in a number of levels, from the very simplest to the highly sophisticated. Welding equipment was originally divided into manual (e.g. when using coated electrodes), semi-automatic (for MIG/MAG welding, with the wire feed being mechanised) and automated (e.g. for submerged arc welding, with both the wire feed and movement of the welding head being mechanised – see Chapter 10).

This third stage is a fully mechanised welding process e.g. using tractor systems. The welding head is fixed on a tractor which is typically mounted on a track driven by a rack and pinion system. As the tractor moves the workpiece is welded (see Chapter 10). The user can adapt tractor systems to suit a particular application, including torch oscillation devices to allow positional GMAW welds to be performed, trailers to carry ancillary equipment (e.g. wire feed units) and tactile seam-following devices. Alternatively, welding can be automated through the use of rotary positioners or welding lathes which can be used to move relatively small components under a fixed welding head. For larger workpieces, column and boom positioners, motorised beams, roller beds and turntables are available.

The introduction of the welding robot provided a further level of mechanisation in that, in addition to starting and stopping, the program-controlled movements of the robot could also include moving between welding positions. Fully automatic welding arrived with pallet-controlled loading and unloading of workpieces, as used in Flexible Manufacturing Systems (FMS). These systems are generally used for smaller and medium-sized items, with the welds being made under shielding gas.

Mechanised welding can improve productivity, the quality of the welds and working conditions. In many situations, mechanisation can be used to perform welding that is not possible with manual processes: an example of this is narrow-gap welding.

15.2 Quality issues in mechanised welding

When an experienced manual welder makes a weld, he or she will notice whether any part of the equipment or process is not operating as it should. He/she monitors the process and makes suitable adjustments to ensure the quality of the final weld. Mechanised welding, on the other hand, faithfully and accurately repeats programmed motions. It is therefore important to ensure that all other factors are kept sufficiently constant to ensure a successful result. The basic principle is that the tolerances that the welding process can accept should be greater than any departures or deviations that are likely to occur. Unfortunately, the acceptable tolerances tend to decrease as pressure to raise productivity increases.

Typical examples of factors that can adversely affect the process are:

- Gap width variations in the welded joint
- The position of the electrode tip relative to the joint

- Problems with wire feed.

The following factors can be important in order to ensure that the necessary quality is maintained:

- Appropriate design and choice of joint type
- Use, wherever possible, of the most favourable welding position
- Preventive maintenance of the equipment
- The use of welding equipment that can control and maintain set values
- The use of monitoring equipment.

Special methods can be used for dealing with difficult situations:

- Weaving, or slow thermal pulsing, can make the process less vulnerable to variations such as gap width changes
- Joint tracking.

Quality standards require welding procedures to be described in a Welding Procedure Specification (WPS). Some form of monitoring should then be employed to ensure that the specified welding parameters are being maintained.

The simplest form of monitoring is the recording of the relevant parameters by monitoring equipment, possibly accompanied by visual display. Typical parameters for such monitoring include current, voltage, wire feed speed, gas flow rate and possibly the travel speed. The equipment should preferably be suitable for connection to a PC for presentation and documentation of the information.

By setting appropriate tolerance limits, the equipment can generate alarms if any of the monitored parameters moves outside the permitted range. The alarms may be generated instantly when a passage occurs, or be made to respond only to an average signal so that brief excursions are ignored. Double limits can also be used, so that traversing the first limit provides a warning, while traversing the second limit stops the equipment.

15.3 Mechanised TIG welding

A good example of mechanisation is mechanised TIG. This produces excellent, smooth welded joints of very high quality. The process is used for applications such as welding stainless steel, nickel alloys and aluminium. Special tools are often used for erection welding of parts such as stainless steel tubes (see Figure 15.1), when welding heat exchanger or condenser tubes into tube plates or for narrow-gap welding (see Page 161).

Equipment for mechanised TIG welding often incorporates advanced control facilities, such as programmable welding parameter 'groups' to suit particular types of joints or materials. Parameters that are usually controlled in this way can include the welding current, pulsing, weaving, wire feed etc.

Tube welding equipment uses special tools as shown in Figure 15.1. When welding horizontal tubes, the welding head must travel round the periphery of the tube, which means that its angle must constantly change to suit the position. It is therefore possible to program different welding parameters for different sections of the weld.

Filler wire is required to fill joint gaps, and is supplied from a wire feed unit. The wire is fed directly into the weld pool, usually in front of the arc and without having first been melted by the heat of the arc.

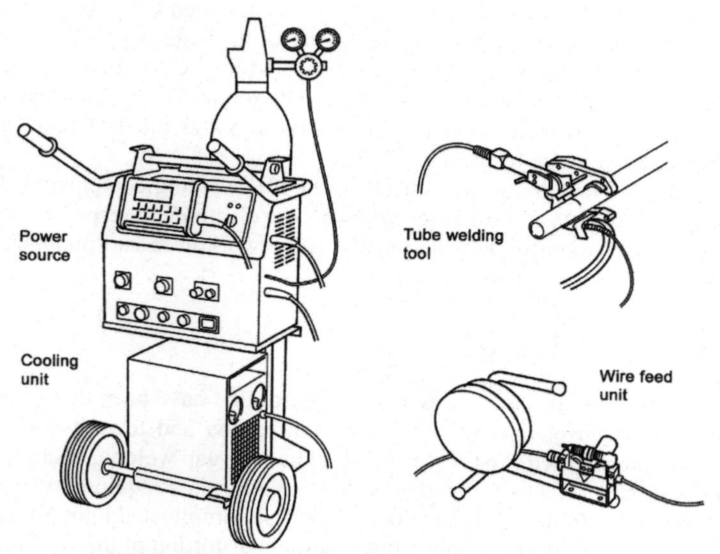

Figure 15.1 Welding equipment for mechanised TIG welding, with the power source and cooling unit mounted on a trolley together with the necessary gas bottle, tube welding tool and filler wire feed unit.

A variant of this is the hot wire system, in which the filler wire is preheated by a separate power supply, as shown in Figure 15.2. The heating current should preferably be adjusted to suit the wire feed speed. The weld speed can be increased, as less of the energy in the arc is needed in order to melt the filler wire material, which results in important benefits when compared with the use of cold filler wire:

- There is less risk of pore formation, as preheating the wire assists the release of gases trapped in the weld pool.
- A high melting rate, up to 8 kg/h of molten material for steel.
- Reduced heat input and less risk of distortion.

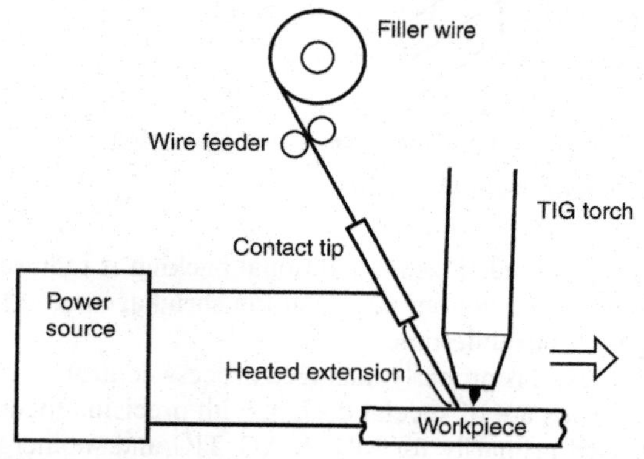

Figure 15.2 TIG welding using the hot wire system. Note that the wire is fed directly into the weld pool without having first been melted by the heat of the arc.

This arrangement requires the welding equipment to be complemented by an additional wire feed unit, together with a special welding gun for the filler wire. The wire is fed into the weld pool at an angle of about 40°, and is heated by AC in order to reduce magnetic arc wind effect on the arc. There is a special TIG torch (TOPTIG) where the wire has only a 20° angle. It is then easier to program the robot as the TOPTIG torch does not need to be rotated through its axis.

In order to keep the arc length constant, mechanised TIG welding equipment often includes Arc Voltage Control (AVC) that raises or lowers the electrode in response to the arc voltage, so that the electrode follows the profile of the workpiece and maintains a constant arc length.

15.4 Narrow-gap welding

The term 'narrow-gap' welding is used to describe processes that have been designed to reduce weld metal volume in butt welds, particularly in carbon and low-alloy steels. These processes are all mechanised. The cost benefits of narrow-gap welding result from its low energy and filler material costs, together with the shorter time required to fill the considerably smaller weld volume. Other factors include less complicated joint preparation, reduced time at elevated working temperatures and less distortion of the workpiece due to the smaller volume of weld metal.

Narrow-gap welding is suitable for joining sheet and plate in the 25–300 mm range. Joints to be welded are prepared with parallel sides or slightly U-shaped, as shown in Figure 15.3. Joints with parallel sides are cheap to prepare, although difficulties can arise as a result of contraction during welding, which has the effect of narrowing the remaining (unwelded) gap.

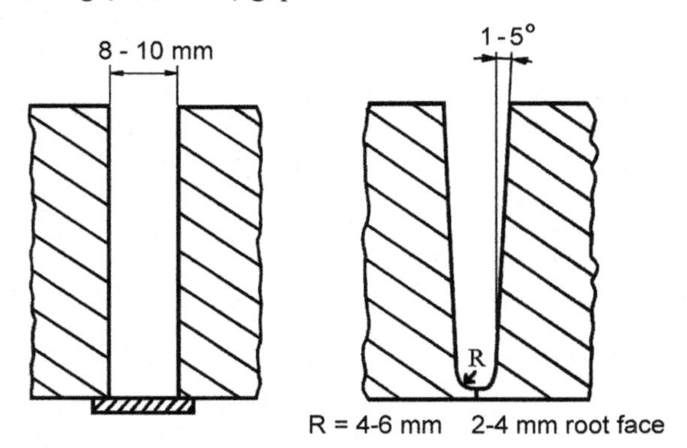

Figure 15.3 Joint types for narrow-gap welding.

The U-shaped joint is to be recommended, as no additional backing is required, it is easier to melt the walls of the joint at the bottom edge and the shrinkage caused as the weld metal contracts on cooling is less troublesome.

A prerequisite for a successful result is properly managed process control, in order to avoid weld defects. The welding heads are of special design, with precision mechanical control. Narrow-gap welding is used primarily for MIG/MAG, TIG and submerged arc welding.

Narrow-gap MIG

To ensure proper melting of the sides of the joint, the arc should alternately be welded to the left and right side of the joint (see Figure 15.4). The arc can also be controlled by winding two filler wires around each other, which causes the arc to move constantly between them, angled outwards and rotating as the filler wire is melted.

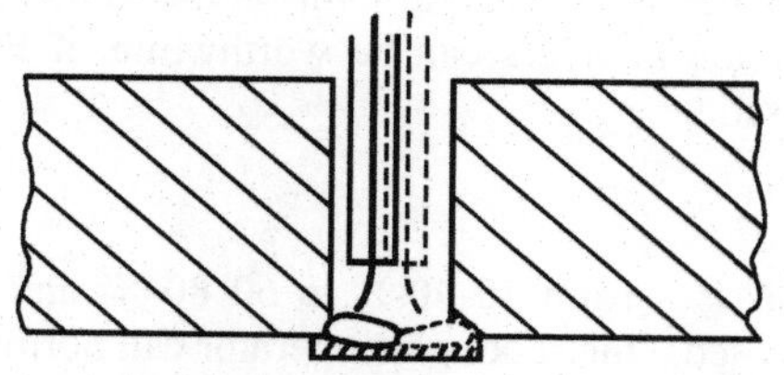

Figure 15.4 MIG welding of two beads beside each other, one at a time, with a rotatable welding head.

Narrow-gap TIG

TIG produces a high quality weld metal in all welding positions. It is particularly suitable for use in narrow joints, bearing in mind its normally otherwise low productivity. The most important elements for successful welding are good gas protection and ensuring that the arc reaches the edges of the joint. This method is used primarily for stainless and low-alloy steels.

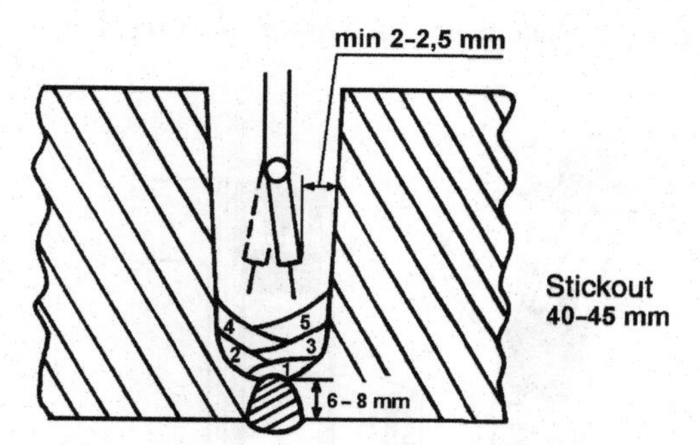

Figure 15.5 A narrow-gap U-joint, suitable for submerged arc welding.

Narrow-gap submerged arc welding

A submerged arc welding head for narrow-gap welding must itself be narrow. The best method of this mode of welding is to lay down beads, alternately to the left and to the right. The electrode should be angled towards the wall of the joint, and be carefully positioned (see Figure 15.5). With the correct welding positions, the slag will loosen by itself. This means that, when welding cylindrical rotating workpieces, the slag falls off by itself. The heat input is low and the welded joint has good strength and toughness.

15.5 Arc welding using robots

The most advanced form of automation is the use of robots. A typical robot cell consists of the robot and a manipulator for the workpiece to ensure that the robot can reach a particular joint and weld from the best position. The robot is programmed to weld from one point to another. In cost terms, a welding robot cell for MIG/MAG welding is economic for a throughput of between 1000 and 60 000 units per year. For numbers less than this, the cost of programming the robot and manufacturing jigs would probably not be recouped. If manufacturing larger numbers of units, it would be better to invest in a larger or more sophisticated installation that would weld several joints at the same time. However, in some cases, even small production numbers can be worthwhile, if FMS equipment or off-line programming can be used.

The benefits of an arc welding robot cell are:

- Higher productivity - the arc time factor rises from 30–40 % to 60–80 %, and the welding speed can also generally be increased. One robot plus operator can normally replace 2–4 manual welders.
- Weld quality becomes more consistent, and generally higher.
- Working conditions are improved, as the operator does not need to stand in the vicinity of the struck arc.
- The necessary organisation and improved control of peripheral activities associated with the introduction of a robot have a beneficial effect on general efficiency.

The drawbacks are:

- A substantial need for training, both for programming and for servicing.
- Closer tolerances on parts to be welded and on jigs.
- A probable need for redesign of parts to make them suitable for robot welding.

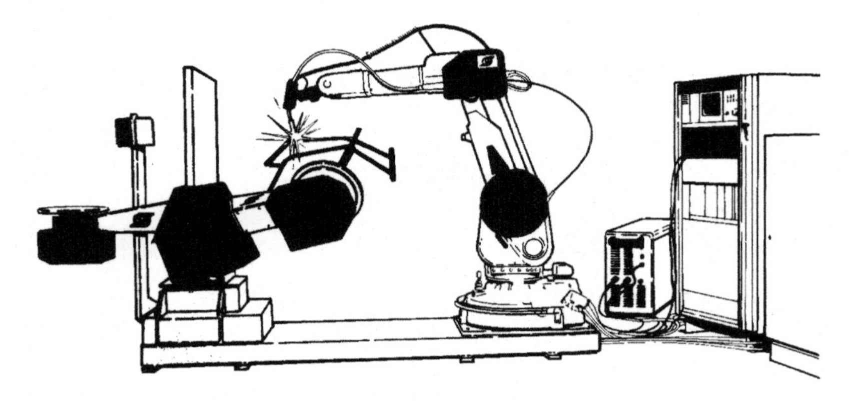

Figure 15.6 A robot cell for arc welding with an interchange type manipulator.

Equipment in arc welding using robots

As noted, an typical robot cell consists of a robot with its control equipment, a manipulator for the workpiece, a welding power unit and other welding equipment. Robots now often have six axes of motion, and there may also be one or more axes on the manipulator that positions the workpiece. The manipulator not only ensures that the robot can reach all the joints, but also to ensure that welds can be carried out in the best position.

Both the robot or the manipulator may execute the welding motion by their own or together. The manipulator may be a two-station interchange type, so that the operator can load and unload jigs while the robot is welding, as shown in Figure 15.6.

There are several ways of increasing the range that robots can cover when dealing with large workpieces, e.g. by mounting the robot on rails or supporting it from a hanging post or a gantry. This means that, in addition to the robot's own six motion axes, with perhaps two on the workpiece manipulator, there can be up to a further three axes for moving the robot. Modern robot control systems can provide full coordination of motion on all the axes, i.e. the robot's own axes and the external axes. An extended working range can also be used to transfer the robot between welding stations so that it can be used with several workpieces.

The robot is programmed to weld from one point to the next on the workpiece. The operator controls the robot with a joystick and identifies the various positions at which an 'event' occurs, e.g. where the robot changes direction, the welding parameters change or the robot starts or stops welding. The coordinates of the positions are saved in the robot's memory, together with the associated instructions. For arc welding, it is not only the position of the electrode that is important, but also the angle of the welding gun relative to the joint. Other welding data such as wire feed speed, voltage and travel speed are included in the programming. If necessary, a weaving action can also be applied to the main travel motion.

Programming is usually carried out on-line at the workpiece, although it is also possible to use special software that allows a considerable part of the work to be carried out on a separate computer. This is referred to as off-line programming (OLP). Off-line programming avoids expensive standstill time for the robot cell, i.e. when it is being used for programming, rather than for revenue-generating work. A large workpiece may require over an hour's welding, and each minute of welding time can take an hour to program.

In its most advanced form, off-line programming uses sophisticated 3D graphics programs into which information on the robot system, jigs and workpiece are loaded, so that the necessary weld motions can be programmed and controlled directly on the screen. Information on the workpiece design can also be provided directly from the Computer Aided Design (CAD) system. Although it may be necessary to perform final fine-tuning on the robot itself, this procedure greatly reduces the standstill time. Other types of off-line programming can be used to convert stored coordinates for a particular model (typical) case: this requires some degree of similarity between the various items to be welded.

Robot installations are typically fenced in order to prevent injuries to persons. Unauthorised persons must not be allowed within the working area of the robot while it is in operation. The normal form of protection is a high fence, with access points protected by light beams that automatically stop the manipulator or robot if anyone attempts to enter the enclosure. Only specially trained programmers or service technicians are allowed to work within the robot's working area, and then only with special protective equipment and procedures. In addition, there are emergency stop buttons, which should preferably be straightforward to reset after operation.

Robot systems may also incorporate maintenance systems. The welding gun can be spatter-cleaned at appropriate intervals, under the control of the robot program. The robot moves to a special position, where the welding torch is mechanically or pneumatically cleaned. There may also be tool change systems, e.g. for replacing the welding gun for maintenance, for changing the welding process or even for changing to a completely different task, such as grinding.

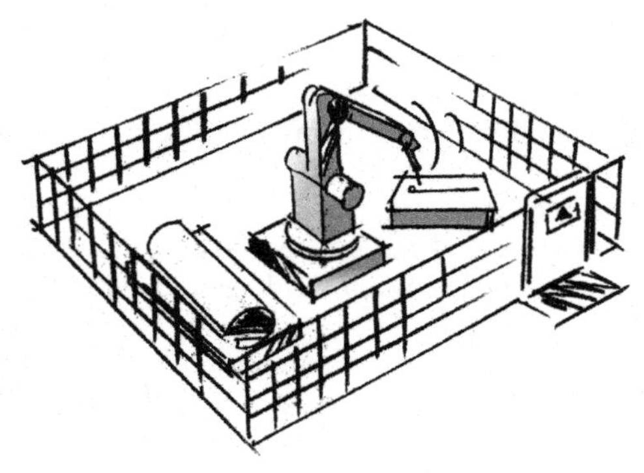

Figure 15.7 A robot welding station inclosed by a fence with an access door.

Welding in flexible manufacturing systems

Flexible manufacturing systems (FMS) process the workpiece from where it may be stored through the welding process and onto its final location in the factory (see Figure 15.8). They can have several positions for loading and unloading the pallets carrying the workpieces. Systems may use one or more robots, and several different types of workpieces can be welded one after the other, with only short changeover times. The use of buffer stocks enables production to continue completely unattended for some time. The entire facility is monitored and controlled by a supervisory computer system.

Figure 15.8 A flexible manufacturing system (FMS) station for robot welding.

Quality issues in robot welding

Spot welding by robots is an established method that has been successfully used since the early 1970s. This is because the required accuracy of positioning spot welds is often

± 1 mm, which was therefore within the accuracy of repeatability of the first generation of robots. The welding clamp is positioned, and then closed to close the gap between the two pieces of metal to be joined and force them into position. Relatively large tolerances can be accepted in respect of the positions of the sheets to be joined.

Arc welding, on the other hand, requires considerably closer tolerances: nevertheless, it is often possible to weld smaller and medium-sized workpieces without the use of joint tracking equipment. The variation in the position of the arc between corresponding workpiece positions must not normally exceed ± 0.5 mm, and achieving this accuracy requires very high accuracy of the workpieces, jigs, manipulators and welding equipment.

The filler wire electrode is the last link in the chain. The wire is seldom completely straight and centred when it emerges from the contact tip. Instead, it is often slightly bent and, if the contact tip is worn, there is a risk of the necessary tolerance requirements not being met. The slight pre-bending of the filler wire is intended to ensure sufficient contact pressure against the contact tip. However, when carrying out robot welding, it is important that this pre-bending is not excessive, and that the contact tips are replaced sufficiently often.

The difference between a human welder and a robot welding several essentially identical items is that the human welder can see if the joint to be welded is too wide or too narrow, and can then adjust the current, weave the welding head, reduce the travel speed, bend the workpiece or take some other steps to ensure a good end result. The robot, on the other hand, may lack the same level of flexibility and might therefore produce a poor welded joint under these conditions.

In most cases, it is possible to find ways of meeting the tolerances in robot welding without having to use joint location or joint tracking equipment. However, when welding large workpieces, it may be impossible to maintain the tolerances required for a successful weld without using such equipment.

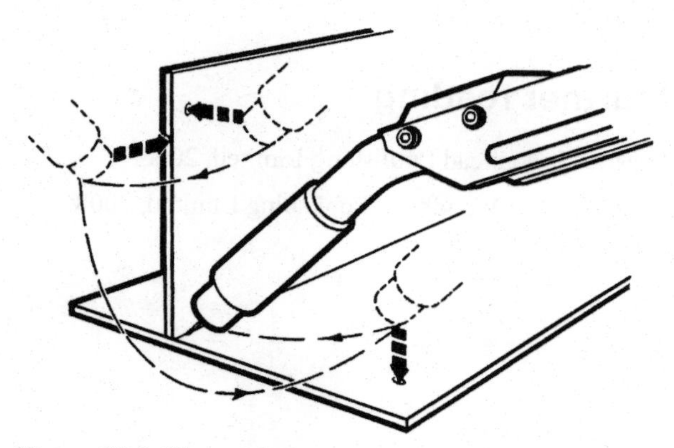

Figure 15.9 The workpiece position is located in order to correct the programmed track so that it coincides with the actual physical position of the joint.

In simple cases, one joint location operation may be sufficient. This is done by the robot before welding starts, by locating the position of the workpiece or of the surfaces of the joint, e.g. by establishing electrical contact between some part of the welding gun and the joint (see Figure 15.9).

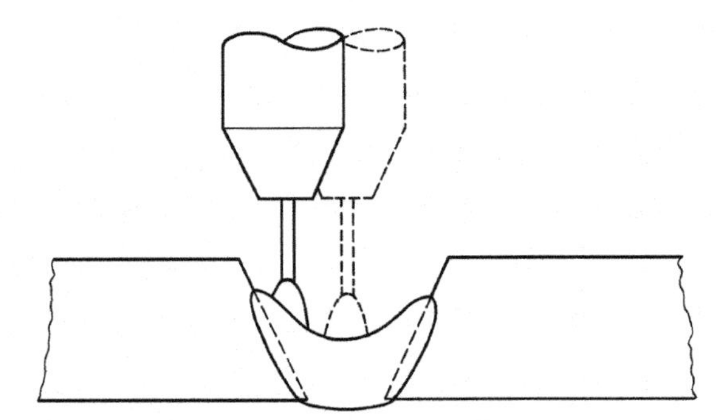

Figure 15.10 If the arc length changes when the electrode weaves in the joint, the corresponding change in the welding voltage can be used for joint tracking.

A location search can be carried out, for example, for three surfaces in mutually perpendicular directions. This is followed by appropriate parallel displacement of the robot's programmed motions, in order to locate the starting point of the weld and eliminate position error, at least in the area where joint location was carried out.

A joint tracking system can be used when making long welds in order to provide real-time correction for any deviations. The commonest system for use with robot welding is to use the arc itself as the sensor (Figure 15.10). The arc is made to weave across the joint, and the control system compares the arc voltage or current at the end positions. A difference is a sign on a deviation between the programmed position and the actual position of the weld. Arc sensing has become popular because it does not require any extra space-demanding equipment, and because it tracks the joint at the position of the arc.

Optical joint tracking devices are also used, e.g. using a laser scanner to determine the positions of the joint surfaces in front of the arc.

15.6 References and further reading

J. Norrish, *Advanced welding processes*, Woodhead Publishing Limited, 2006.

Y. Zhang, *Real-time weld process monitoring*, Woodhead Publishing Limited, 2008.

16 Soldering and brazing

16.1 Introduction

Both soldering (which can also be called soft soldering) and brazing (which is also referred to as hard soldering) are processes that rely on the penetration of a capillary gap by a molten filler material at a temperature which is below the melting point of the materials to be joined. The distinction between the two processes depends on the working temperature:

- if it is below 450°C, the process is referred to as soldering.
- if it is above 450°C, it is referred to as brazing.

As the working temperature is always lower than the melting temperature of the base material, and generally very much lower, the base material will remain solid throughout the process. This is the main difference in principle between soldering/brazing and welding.

Soldering and brazing are important methods of bonding metals used particularly in mass production. They are used to join materials and components for which welding is unsuitable due to the considerably higher temperatures and limited abilities to bond different metals. Soldering and brazing are suitable bonding processes for joining ferrous metals to non-ferrous metals, and for bonding metals having very different melting temperatures. They are the main methods of bonding used in the manufacture of products made of copper or copper alloys. The processes are economic due to the lower working temperatures and to the fact that there is generally no need for any substantial chipping, grinding or cleaning of the joint as a finishing process.

The choice of soldering method depends on factors such as the design of the workpiece and the quantities to be soldered. We can distinguish between various types of soldering and brazing technique:

- *Manual soldering*: a technique using a soldering iron, which is usually electrically heated. This method is used only for soldering.
- *Flame heating*: a soldering and brazing technique using an open flame.
- *Solder baths*: a method of soldering using a bath containing molten solder and in which the parts may be completely or partially immersed.
- *Wave soldering*: a technique used when electronic components are soldered to a printed circuit board.
- *Furnace soldering*: a method in which the workpiece, with pre-placed solder (known as solder preforms) and flux, is heated in a furnace, generally on a conveyor. Parts may be protected by an inert atmosphere.
- *Induction heating*: a technique in which the heat is generated by high-frequency induction.
- *Resistance heating*: a method where electric current is passed through the workpiece to produce resistive heating. As with furnace soldering, solder and flux can be pre-placed.

- *Braze welding:* a method similar to gas welding where filler metal is not distributed by capillary action. It is typically used for fillet joints, V or X joint.
- *Arc brazing*: a technique using an electric arc and brazing metal. This method is similar to MIG (*MIG brazing*) or plasma welding.
- *Infrared soldering*: a method which uses infra-red radiation from high-power halogen lamps and reflectors.
- *Laser soldering:* a technique which uses radiant heating.

Soldering or brazing involves heating the area to be joined to the working temperature of the filler metal, or to a somewhat higher temperature. A flux is generally used in order to chemically remove oxides from the surfaces and to prevent new oxide layers forming during the heating process. If the surfaces are sufficiently clean, the molten filler metal wets (spreads over) them and diffuses into the base metal. This produces an alloy of the filler and base material in a thin layer in the bond zones (see Figure 16.1), thus producing an uninterrupted metallic bond in the form of the soldered or brazed joint. Thermodynamic processes result in both materials diffusing into each other: the elevated temperature of the process causes elements from the added metal to diffuse into the base material and vice versa.

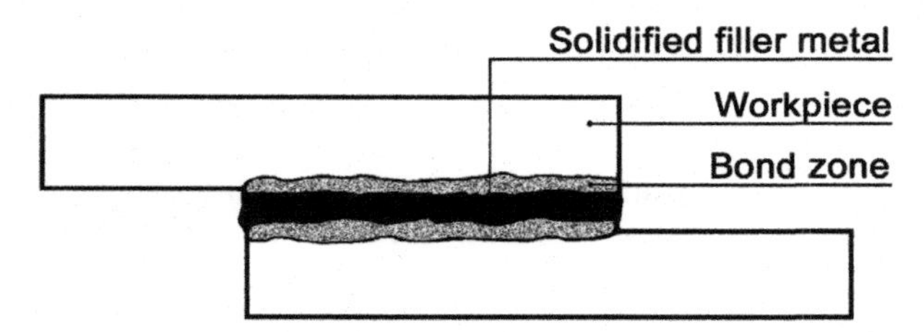

Figure 16.1 Schematic diagram of a brazed joint.

The composition and structure of the bonding layer is critical to the strength of the joint, and so it is important to choose a filler metal that is compatible with the base material.

The solidified layer of filler metal in the joint must be thin if the best strength and fill properties are to be achieved, which means that only in exceptional cases may the gap between the two pieces to be joined exceed 0.5 mm. Narrow gaps draw the molten metal into them by capillary attraction, with the best fill being obtained when the gap is between 0.05 mm and 0.25 mm. A brazing alloy with a narrow working temperature range normally penetrates better than one with a wider working temperature range.

16.2 Soldering

Copper and its alloys are the materials that are most commonly joined using soft soldering. The most commonly used solders for soft soldering are alloys of tin and lead, although a wide range of other alloys are used to a lesser extent, often for special purposes. Table 16.1 lists the compositions of normal types of solder.

Solders used for aluminium, for example, include alloys of zinc and cadmium, or zinc and aluminium. Solder is often supplied in the form of wire or bars, generally with flux cores.

Safety regulations (the RoHS Directive) restrict the use of lead, cadmium and other elements in solder due to their toxicity. Where necessary, lead in solder must be replaced by lead-free alternatives. This therefore raises the melting point of the solder.

Fluxes are often weak acids or salts. As they are corrosive, their residues must be removed from the joint after soldering. An alternative, which does not normally result in corrosion if traces are left, is the use of resins dissolved in an organic solvent. There are also liquid and paste fluxes.

Strongly active fluxes must be used for soft soldering of materials with chemically stable oxide films, such as aluminium or stainless steel, in order to remove the oxide and allow the solder to come into contact with a sufficiently clean metal surface. To produce a satisfactory joint, stainless steel should be pickled immediately before soldering.

TABLE 16.1 Composition of common solders.

Sn	Pb	Cd	Bi	Ag	Sb	Zn	Cu	Melting range, °C	Application
99.9	Trace							232	Tinning, (food tins)
63	37							183	Wave soldering and solder baths
60	40							183-188	
50	50							183-226	Brass, solder baths
42			58					145	Heat-sensitive parts
60	38						2	183-188	Electronic assem-blies, corrosion-resistant
22	28		50					95-107	
12.5	25	12.5	50					70-74	
96.5				3.5				220	Cu pipes (HVAC work)
95.5				3.8			0.7	217	Electronic items (lead-free)
97.5				2.5				221-225	
95					5			233-240	
	97.5			2.5				305	Electrical work at high operating tem-peratures
95					5			233-240	Food industry (not copper)
		70				30		300-350	Al and Al alloys

Although much soft soldering is carried out manually, the process is also very suitable for mechanisation. In the latter case, the gap between surfaces to be joined should be narrow, not more than 0.2 mm, in order to ensure that the molten solder can reach all parts by capillary attraction. The commonest joints for such processes are various types of overlap joints.

16.3 Brazing

Brazing can be performed manually, mechanised or automated. The key factors in determining the quality of a brazed joint are the method of brazing, the composition of the brazing filler metal and the flux. Joint surfaces must be carefully cleaned if the best bond quality is to be obtained. Oxides must also be reduced, and/or oxide formation during heating must be prevented. One way of doing this is to make the joint in an air-free environment, e.g. in a furnace with a suitable inert gas atmosphere, or under vacuum. However, brazing is generally carried out in air, which means that self-fluxing filler metals (copper-phosphorus) or a separate flux must be used.
The flux consists of a mixture of various metallic salts, and is applied to the joint surfaces as a high-viscosity liquid, as a paste or as a powder. Heating drives off the carrier (often water), before the salts melt and react with the oxides on the joint surfaces. The choice of flux depends on the type of workpiece material and the brazing temperature of the filler metal.

For use as a filler material, a metal or alloy must have a lower melting point than the workpiece material, and must be able to wet the workpiece material. In addition, when molten, it must flow sufficiently easily to enable it properly to fill the joint. The resulting alloy must provide the necessary mechanical and physical properties, while the filler metal must not be vaporised to any significant extent during the heating process, as this could result in poor fusion. Metals in the following groups are those most commonly used as brazing filler metals: see also Table 16.2.

Silver brazing alloys consist of alloys of silver, copper, zinc and sometimes also cadmium. They flow easily, with low working temperatures of 600–800°C. They can be used with all brazing methods, and for almost all materials except aluminium and magnesium alloys. However, due to the toxicity of cadmium, filler metals containing this substance must not be used unless full evacuation of brazing fumes is provided.

Copper phosphorus filler metals are based on alloys of copper and phosphorus, and may also contain silver. They are used almost exclusively for brazing copper and copper alloys. As phosphorus reduces copper oxide, the metal is self-fluxing, and so no additional flux is normally needed. These filler metals are not suitable for brazing steel or nickel, as phosphides can be formed and embrittle the joint zone.

Copper and brass. Oxygen-free electrolytic copper is used for furnace brazing. Brass (from which the name 'brazing' is derived), when used as a brazing filler metal, consists mainly of copper (40–80 %) and zinc, and may often include smaller quantities of tin and silicon. It is used for brazing materials such as mild steel.

High-temperature brazing materials consist of nickel-based alloys and silver-manganese alloys. They are used for brazing components intended to operate at temperatures well above normal ambient temperatures, such as parts for gas turbines or steam turbines.

Aluminium brazing metals. Aluminium and its alloys are brazed with filler metals consisting of aluminium-silicon alloys or aluminium-silicon-copper alloys.

Solders are available in different forms, such as wire, bars, strip, foil and grains. Solder can also be supplied in preform shapes such as rings, washers etc. Some solders in bar or wire form may be delivered coated or cored with flux. Solders are also available in paste form, with flux already included for mechanised production processes.

TABLE 16.2 Composition of common brazing filler metals.

Type of alloy	Ag %	Cu %	Zn %	Sn %	P %	Mn %	Ni %	Al %	Si %	Fe %	Melting range, °C	Working tempera-ture, °C
Silver brazing alloy	55	21	22	2							630-660	650
	45	28	25	2							640-680	670
	49	27.5	20.5			2.5	0.5				670-690	690
	44	30	26								675-735	725
	30	36	32	2							665-755	745
	18	45.8	36						0.2		690-810	810
Copper-phosphorus alloys	5	89			6						650-810	715
	2	91.8			6.2						650-810	710
	15	80			5						650-800	705
		94			6						710-850	850
Bronze / brass		48	41.8				10		0.2		880-910	900
		47.5	41.5	0.1		0.3	10		0.3	0.3	890-900	900
Aluminium alloys								88	12		575-590	580

Types of joints in brazing

Brazed joints are generally some form of overlap joint. To a lesser extent, joints may be of butt type, but should be chamfered to improve the joint strength. The gap width should be in the range 0.05–0.5 mm, and preferably 0.1–0.2 mm.

The strength of a brazed joint depends on various factors, including the area of the joint. The greater the area, the greater the forces that the joint can withstand. As it is not necessary, when preparing a brazed joint, for the joint to be manually accessible throughout its length, a geometry such as that of the butt overlap joint or similar can often be used (see Figure 16.2).

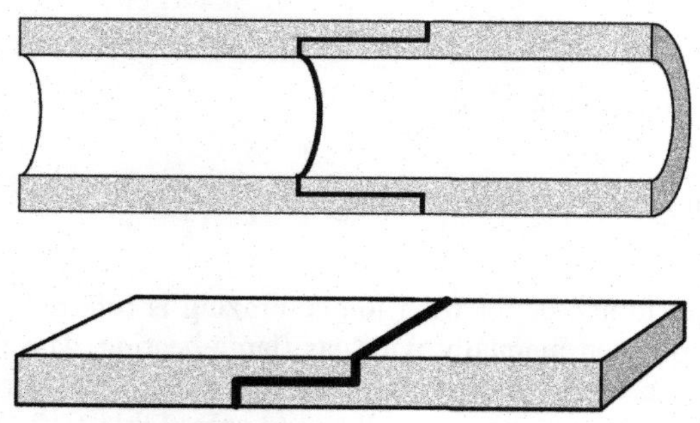

Figure 16.2 Butt overlap joints.

An overlap width equal to about three times the thickness of the thinner part should be aimed at, in order to provide sufficient strength of the joint: see Figure 16.3.

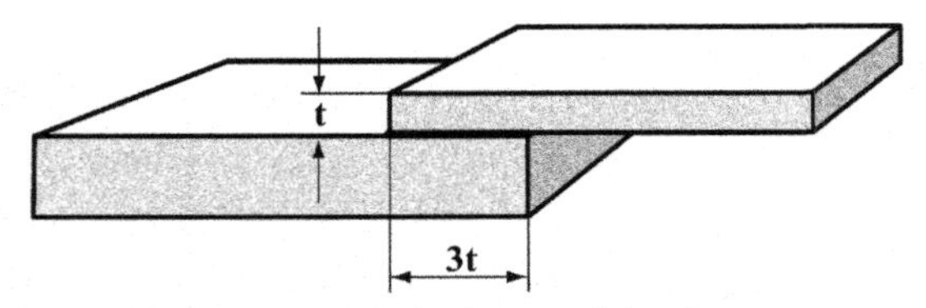

Figure 16.3 Three times the thickness of the thinner part is a suitable lap length.

A particular problem when making brazed joints is that of gap width. When joining dissimilar metals and/or different thicknesses, the width of the gap changes (either increasing or decreasing) as a result of different rates of expansion of the metal as it is heated to the working temperature (see Figure 16.4). A suitable gap width is normally obtained by placing the parts in contact with each other, without any additional pressure.

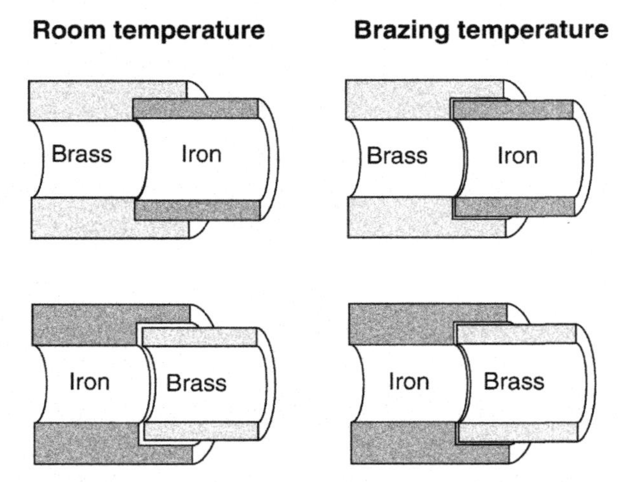

Figure 16.4 Different combinations of materials can result in the gap increasing or decreasing as the parts are raised to working temperature.

Brazing methods

This section discusses a number of common brazing methods:

- Braze welding
- Arc brazing (including MIG, TIG and Plasma arc brazing)
- Laser beam brazing

When used to make V or X butt joints, or for fillet joints, brazing is referred to as braze welding, and is generally performed manually with gas flame heating. The filler metal must be relatively viscous, in order to provide the best fill of the joint. Various types of brass alloys and appropriate fluxes are generally used: see Table 16.2. This process is now less common, but is still used for making joints in copper and copper alloy pipes, as well as for brazing bronze and cast iron.

Arc brazing is the name given to brazing methods based on arc welding methods as MIG, Plasma and TIG process. It is used similar to braze welding, i.e. the process is less dependent on the capillary force between the joint surfaces, than on the strength of the weld metal that fills the joint. Applications include automotive body components manufactured of zinc coated steel sheet. In arc brazing heat input is reduced and there is less vaporisation of zinc. It is therefore an alternative to conventional fusion welding which faces problems due to the burn-off of the zinc coating and the occurrence of process instabilities and spatter. Brazing speed can often reach the double value compared to welding and there is less need for post-processing treatment of the brazed seam. Filler material can be low-alloy copper-based (e.g. CuSi3Mn1), aluminium-bronze or tin-bronze based.

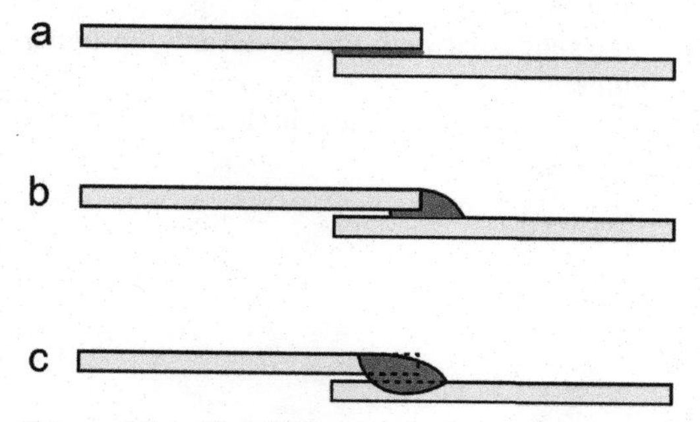

Figure 16.5 The difference between brazed and welded joints: (a) capillary brazing, (b) arc brazing, (c) welding.

MIG brazing involves replacing the electrode in an MIG welding torch with an electrode made of a copper based brazing filler metal. The power ratings are set so that the filler metal melts, but the edges of the joint are only heated, and not melted, by the arc. The shielding gas used is pure argon or argon with a small amount of active gas that improves the arc stability. It is common to use argon with up to 2 % oxygen or CO_2 as shielding gas. Pulsed MIG arc can be used to improve arc stability. MIG brazing allows a high joining rate. Brazing speeds up to 3 m/min can be achieved.

TIG brazing uses a mechanised feeding of the brazing filler wire. Pure argon is used as shielding gas.

Plasma arc brazing uses a plasma torch combined with a wire feed of brazing metal. The plasma arc is more stable and gives better brazing properties compared to MIG brazing. Generally argon is used both as plasma and shielding gas but small additions of other gases (He, H_2) could be used to improve the brazing speed.

Laser beam brazing is an excellent alternative to the arc brazing technique when joining thin sheets of steel. Both CO_2 and Nd:YAG lasers are used but also high power diode lasers are becoming available. For brazing the focus of the beam does not need to be as precise as for welding. The ability to increase the laser beam spot size by defocusing is an advantage. Filler wire is supplied by a wire feeder but can also be preplaced. Laser beam brazing cause very small heat-affected zones. The seam becomes very smooth and allows a minimum of after-treatment before painting.

Definitions

Soldering, Brazing. Bonding metal parts by means of a molten bonding material (*solder, brazing filler metal*) that has a lower melting point than the base material(s), and which wets the surfaces of the joint.

Solder, Brazing filler metal. The metallic material used to make the soldered or brazed joint.

Flux. A meltable or gaseous material, intended to dissolve oxides and to prevent the further formation of oxides.

Wetting. The ability of the solder or brazing metal, when molten, to spread over and connect the joint surfaces (the substrate).

Diffusion. The alloying process, with one of the components remaining in the solid state throughout.

Working temperature. The temperature to which the joint must be raised for the solder or brazing alloy to melt and wet the joint surfaces.

Melting interval, Solidifying interval. The range between the lower and the upper melting temperature (solidus and liquidus temperatures).

Capillary joint. The solder is drawn into the joint – which must be less than 0.5 mm wide – by capillary attraction.

17 The welding environment and welding safety

17.1 Introduction

Welders are exposed to a number of different environmental problems. These include fumes and gases, as well as ultraviolet radiation from the arc. Perhaps the commonest problem is that of minor burns from radiation of heat during welding or from a loose drop of molten metal. It can also be difficult to work during the summer, as welders need to wear full protective clothing.

Short arc welding, for example, is relatively noisy, producing noise levels up to 80 dB. Welding is often also accompanied by noisy grinding work in the vicinity. By using a suitable welding process, shielding gas and welding technique that minimises grinding and slag removal, noise problems can be reduced.

In some cases, the spatter produced by welding can cause discomfort and even burns. The risks increase in connection with overhead welding or if the welding is performed in confined spaces. Spatter can be reduced by using the correct welding parameters and an appropriate shielding gas with a high argon content. Fine spatter is normally fairly harmless. To avoid these problems, it is important to use fully-fitting clothing and clothing made of suitable heat-resistant material.

17.2 Welding fumes and gases

Fumes and gases from the use of coated electrodes are produced mainly from the electrode coating and core material. Welding fumes include small solid particles of iron oxides and manganese oxides, while the coating produces mostly what is known as inert dust, although basic electrodes will also release fluorides. Evolution of smoke and fumes increases with increasing welding current, with up to 1 g/minute being evolved from the thickest high-yield electrodes.

Manganese, which is often a constituent of the smoke, can harm the nervous system, known as manganism. Welding high-alloy steels and stainless steels releases chromium and nickel compounds. A low hygienic limit is set for this fume, as it is a cancer risk.

Materials with various surface treatments also require particular attention. Welding galvanised steel produces substantial quantities of fume due to the low boiling point of the zinc. Inhalation of this fume can cause metal fume fever (zinc shakes) with nausea, while red lead or other paints that can release irritants must be removed from weld areas before welding. Polyurethane paints and foamed insulation form isocyanates when heated, which can affect the airways and possibly cause asthma. Oil fumes are formed if the workpiece is oily or greasy.

Most harmful substances have occupational exposure limits (OEL) which are regularly revised. The most common of these limit values specifies the average concentration which does not normally represent a health risk during eight hours of work a day (level limit value). A maximum exposure limit or short-term limit value is also specified for certain substances (see Table 17.1).

As the gases and particles which form affect the body in different ways, it is important that the regulations issued by the authorities and the instructions issued by manufacturers are followed in order to avoid ill-health. Material Safety Data Sheets (MSDS) are also available. On these sheets, the manufacturer provides detailed information. Hygienic limit values specify the maximum concentration of some contaminant that is regarded as not presenting a health risk. They are expressed as average values.

Some abbreviations: MAC = Maximum Admissible Concentration. TLV = Threshold Limit Value, OEL = Occupational Exposure Limit.

TABLE 17.1 Permissible limit values (2007) for some constituents of welding fumes.

Substance	Threshold limit value, mg/m^3
Dust, inorganic, respirable	5
Iron oxides	3.5
Manganese, total	0.2
Manganese, respirable dust	0.1
Fluorides	2
Chromium (VI) compounds	0.005
Nickel compounds	0.1

The normal protective measures consist of good ventilation, preferably in the form of local extraction immediately above the weld. The ideal is to arrange some form of spot extraction that captures the fume at source. Various systems are available, depending on how mobile the welder needs to be. A common arrangement is that of adjustable fume extraction arms which can be positioned close to the welding operation.

On average, the fume produced by MIG/MAG welding is less than that produced by the use of coated electrodes. Particular care should be taken in certain cases:

- If unusually large amounts of fume are produced. Cored wire containing flux, and welding with high welding data, can produce substantial quantities of fume.
- Ozone is formed for example when MIG welding aluminium at high currents and with a high-radiance arc. Note, however, that shielding gases are available that actively help to break down the ozone.
- The fume from aluminium can cause damage to the nervous system.
- It is important to avoid welding in the presence of chlorinated hydrocarbon solvents (e.g. trichloroethylene): a chemical reaction can produce phosgene, which is poisonous and damages the lungs.
- The use of CO_2 as the shielding gas can produce carbon monoxide which, under certain circumstances, can reach hazardous levels.
- To avoid fumes and gases from paint or other surface treatment, the base metal must be cleaned for at least 10 cm, sometimes even more, from the point of welding.

In these situations, and also when welding in confined spaces where there is insufficient ventilation, it can be appropriate to use a fresh air breathing mask (Figure 17.1). Welding guns with integral extraction can also remove most of the fume before it reaches the surrounding air (Figure 17.2). The position of the extraction nozzle should be adjustable, in order to avoid interfering with the shielding gas.

Figure 17.1 When welding in confined spaces where there is a risk that the concentration of fumes and gases could be too high, the welder must use breathing protection with a supply of clean air.

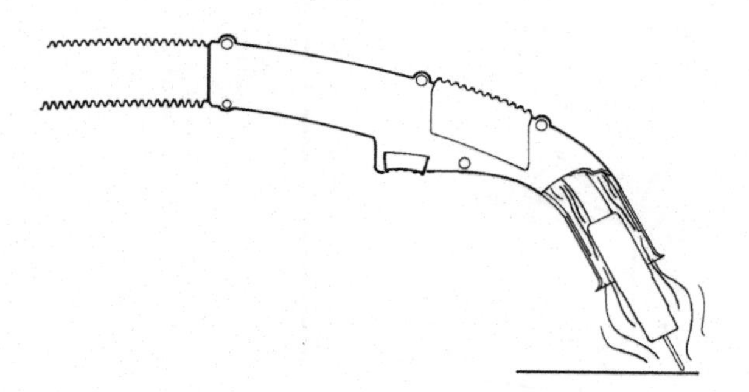

Figure 17.2 Welding gun with integral fume extraction.

17.3 Electrical hazards

Arc welding is unavoidably associated with the risk of contact with live parts. It is therefore important to be acquainted with the risks and to know how to avoid them. Human beings are extremely sensitive to current that passes through their bodies: serious physical injury can be caused by currents of just 20 or 30 mA. Physical injuries, such as falling from scaffolding or a ladder, can result indirectly from very low currents as a result of a sudden and involuntary reaction to an electric shock. The choice of the type of current (alternating or direct) during arc welding is important, as the risks associated with alternating current (AC) are as much as four times greater than those associated with direct current (DC). We are most sensitive in the frequency range of 15-100 Hz.
The effects that result from current passing through the human body and eventually the heart depend on:

- the level and duration of the current,
- the current path through the body,
- the frequency of the current.

The level of the current depends on the open circuit voltage of the power source and also on the resistance of the body. The resistance varies, depending on the size of the contact surface of the live parts, the presence of moisture and the quality of protective clothes and gloves the welder is using. Direct current is an advantage when welding in a confined space where the risks are greater (such as moist, hot or confined spaces with conductive parts or walls). The risk that the operator will get an electric shock is increased when he works in such spaces.

Damage to material or equipment may occur if the return cable is connected inappropriately. It should be connected as close to the welding point as possible. Stray currents through crane cables or lifting hooks can cause partial or complete annealing (see Figure 17.3). In the same way, damage can be caused to the bearings of rotating machines if sufficiently high currents pass through them. The size of welding cable needs to be chosen to suit the highest welding current used, and the cable connection must be fixed in such a way as to ensure that it cannot come loose.

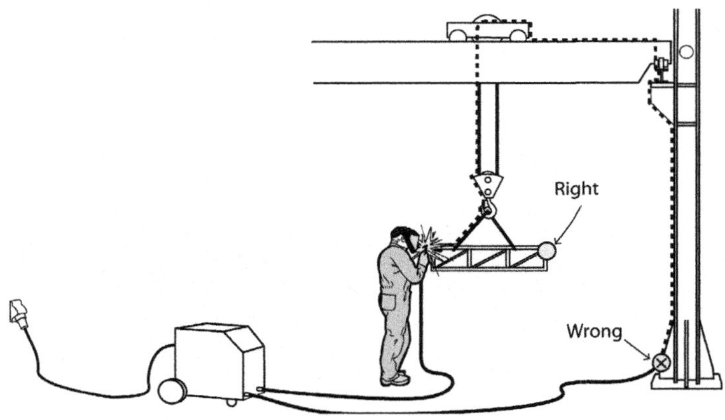

Figure 17.3 To avoid damage to material or equipment the return cable should be connected close to the welding point.

It is not only weld spatter that can cause fires. Heating caused by the welding current also presents a risk. If, for example, the return conductor is not properly connected, the return current may take a different path, with resulting overheating and ignition at unexpected positions. If the welding current returns via the wrong paths, it can also cause damage to mains cables and earth connections, as they are not of a size to carry such currents. Damage can also occur if the workpiece is earthed and there is a fault in the welding current circuit.

Electromagnetic fields

The magnetic field close to a welding current cable is relatively high: see Figure 17.4, which can in some cases risk exceeding the EU permitted limit values. The diagram shows the flux density as a function of distance from a welding current cable carrying

100 A. Alternating current or varying direct current creates magnetic fields which in turn can induce currents in the human body. Physiological reactions can arise when these currents are of the same order of size as those in the body's own signalling system, e.g. in nerves. Sensitivity also increases with increasing frequency: the most harmful in this respect is therefore AC welding, or DC welding with rapid pulsations. Conditions are particularly difficult if the welding current is high and the supply and return current cables are not close together. If they are close, the magnetic fields from each cable largely cancel each other out.

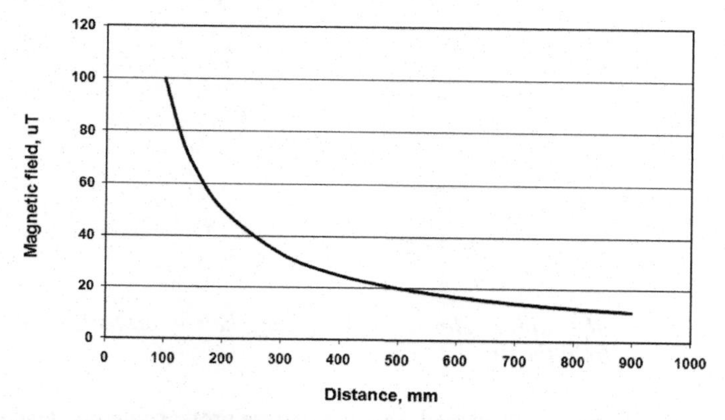

Figure 17.4 Magnetic field strength falls off rapidly with distance. The diagram shows the flux density as a function of distance from a welding current cable carrying 100 A.

Power units can also cause magnetic fields, and particularly those which use a magnetic leakage core to adjust the current. The following simple safety measures should be applied:

- Keep supply and return current cables close together as far as practically possible.
- Don't wind welding cables around the body.
- Use a support or counterbalance arm to keep the cables away from the body.
- Welding with DC is preferable to welding with AC.
- Mechanisation, e.g. with robots, improves working conditions in several ways.
- Anyone with a cardiac pacemaker should be particularly careful. Discuss conditions with a doctor first.

17.4 Arc radiation

The arc is a strong source of radiation in the infrared, visible and ultraviolet ranges of the spectrum. Arc eye (Flash) is a strongly irritating inflammation of the cornea of the eye, caused by ultraviolet radiation. It takes the form of pain and irritation, starting about six hours after exposure. However, it normally subsides naturally, and has no permanent effects. Special protective glasses must be used for the eyes, and all skin should be protected by fully-covered clothing, and be complemented by suitable gloves (see Figure 17.5).

Figure 17.5 Clothing with a leather apron, gloves and a welding helmet protects from heat, spatter and UV radiation from the arc.

A *welding screen* or welding helmet with special protective glass protects against visual radiation is needed (Figures 17.6 and 17.7). As an alternative to ordinary protective glass filters, there are welding helmets with liquid crystal screens that senses the arc light and switch rapidly between the clear and dark position.

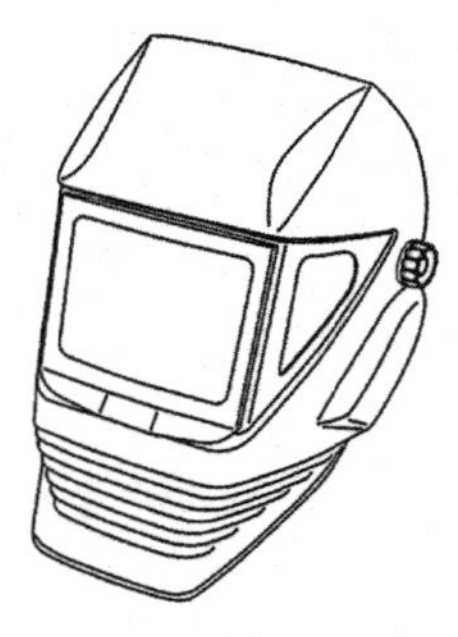

Figure 17.6 A welder's helmet with automatic darkening window, which makes it easier to position the electrode prior to striking the arc.

17.5 Ergonomics

Welding thicker, heavier and/or larger parts manually and during assembly welding involves more static loading on the welder. The welding times are longer and the weight of the equipment is greater. In addition, the working position in this case is dependent on the position of the weld joint. Working with the hands in a high position or above shoulder level should be avoided whenever possible. Overhead welding is unsuitable from an ergonomic angle.

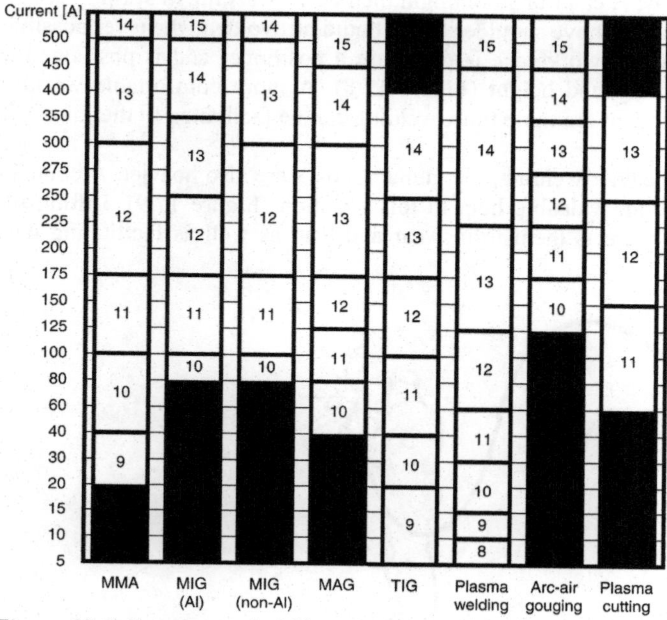

Figure 17.7 Recommended filter shade levels for protective glass filters. More detailed information is given in the standard EN 169.

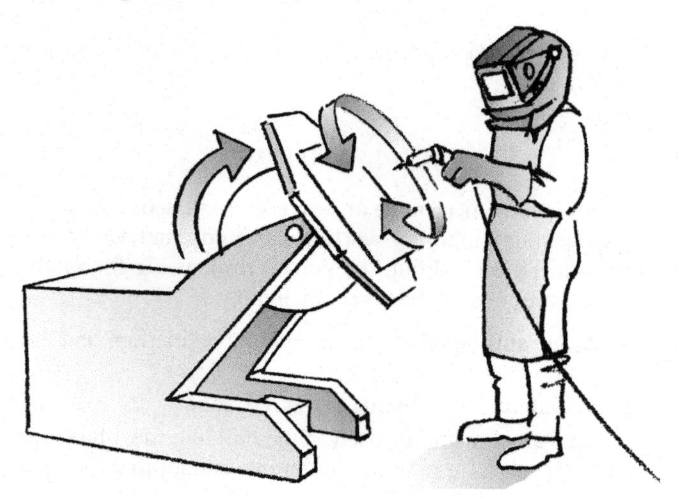

Figure 17.8 A positioner for the workpiece.

Welding small items in fixtures is often characterised by many short welds, with monotonous, unchanging movements between them. When planning a workplace, the working height plays an important part in creating the correct working position. In this context, positioners and lifting tables can be very useful. The working position is partly determined by the welder's need to have his/her eyes close to the workpiece to be able to see the molten pool clearly while welding. If the working height is too low, the welder has to bend to see properly. A chair or stool might then be very useful. Working with the hands in a high position at or above shoulder level should be avoided whenever possible. It is also a good thing if the workpiece is placed in a positioner and is positioned to ensure the best accessibility and height (Figure 17.8). A more comfortable working position can be created and, at the same time, welding can be facilitated as the joint is in the best welding position.

In conjunction with heavier welding, the gun and hoses are also heavier. A counter-balance support can provide valuable help in this situation (Figure 17.9). Lifting the hoses off the floor also protects them from wear and tear, as well as facilitating wire feed.

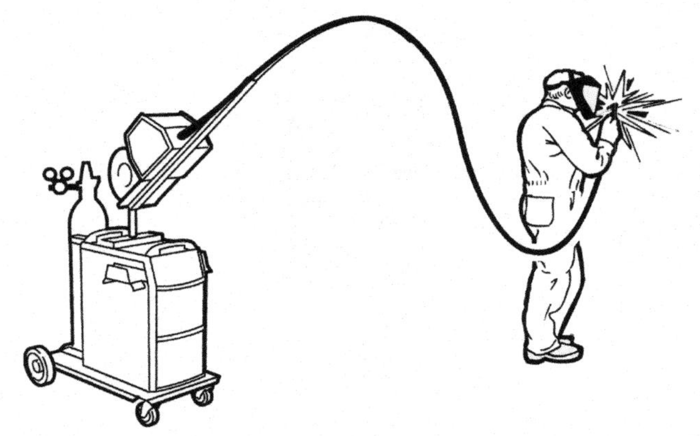

Figure 17.9 A counterbalance arm reduces the weight of the hose bundle over the entire working area.

17.6 Fire risks

Welding and thermal cutting is an obvious source of fire risks. Experience shows that the risks are greatest when undertaking temporary work in areas not intended for welding operations. If such work has to be carried out, the person responsible for safety must decide on what protective measures need to be taken, including:

- Making sure that the staffs are sufficiently familiar with the regulations and the ways risks can be avoided.
- Cleaning and removal of combustible materials in the risk zone.
- Any holes or gaps in combustible materials used in the building must be covered or sealed so that weld spatter or sparks, e.g. from gas cutting, cannot find their way in.
- Using water to dampen the area before and perhaps after work.
- Screening off the area.
- Ensuring that adequate extinguishing equipment is available.
- Monitoring and after-inspection (one hour) by a fire guard.
- If welding or cutting has to be performed in places where there is a risk of fire, a safety officer should assess the preventive action that needs to be taken.

Figure 17.10 Risks are greatest in connection with temporary work in premises or areas not normally intended for welding.

Fires caused by welding and cutting are largely due to a lack of knowledge, carelessness and insufficient protection. Training and effective protection programmes are essential in this context. Workplaces where the risk of fire is high are particularly dangerous welding sites, e.g.:

- Places where flammable substances, such as petrol and oil, or flammable gases are handled.
- Premises containing packaging material or timber goods.
- Construction sites where sparks can more easily spread into areas that are not readily accessible, such as walls made of wood or containing flammable insulation.

When working in dangerous environments of this kind, welding inspections must always be conducted. In some cases, the approval of the municipal safety officer and insurance companies must be obtained.

At other places, where the material is perhaps less flammable, ignition can begin with slow combustion. It may then develop into a fire with open flames. A relatively long period can pass before the fire is discovered and, if the seat of the fire is also difficult to reach, extinguishing the fire is more difficult, even if fire-fighting equipment is available.

Development of fire in different environments

Fires in PVC plastic, which is often found in electrical cables or other interior design materials, generate hydrogen chloride vapour which, together with the moisture in the air, creates hydrochloric acid. This is a powerful irritant and it is also highly corrosive when it comes to metals. In addition, it can damage sensitive electronic equipment.

There may be a risk of explosion when flammable substances such as petrol, oil or paraffin are heated. If they do not ignite directly, there is a real risk of explosion as they first vaporise.

Remember that the heat produced by welding or cutting a pipe can be conducted into a nearby wall and cause a fire, even if the temperature is relatively low. A sufficiently large spark from welding, cutting or even grinding could cause a fire, even if it is not red-hot.

Fires in enclosed spaces such as walls and insulation can develop relatively slowly. Always call the fire brigade if a fire of this kind occurs.

Call the fire brigade even if it appears that the fire has been put out, as it is important to check that no source of slow combustion is left.

17.7 References and further reading

J. Blunt and N. Balchin, *Health and safety in welding and allied processes* (Fifth edition), Woodhead Publishing, 2002.

Welding and cutting - risks and measures. ESAB AB, Sweden.

18 Welding residual stress and distortion

18.1 Introduction

If a material is subjected to localised heating (such as welding), it is not heated uniformly. The metal in the weld region will always be hotter than the surrounding metal, and it will expand and contract much more. This creates potential stresses and distortions that can cause failure.

18.2 Residual stress

If a strip of metal is heated and allowed to contract, it will shrink back to its original size. However, if the metal is restrained in some way, a tensile stress is introduced as the metal cools and contracts. This is known as residual stress. In the welding process, a pool of molten weld metal cools down and shrinks. However, the cooler base material acts as a restraint on the contraction of the weld pool, creating residual stress. In many cases the pieces to be welded may be held in position (e.g. by clamps) which introduce additional stresses as the weld cools. Residual stress is thus caused by uneven expansion and contraction of a welded joint coupled with restraint. This stress can cause failure, for example if there is a defect in the stressed area. Thermal cutting produces residual stresses similar to those caused by welding.

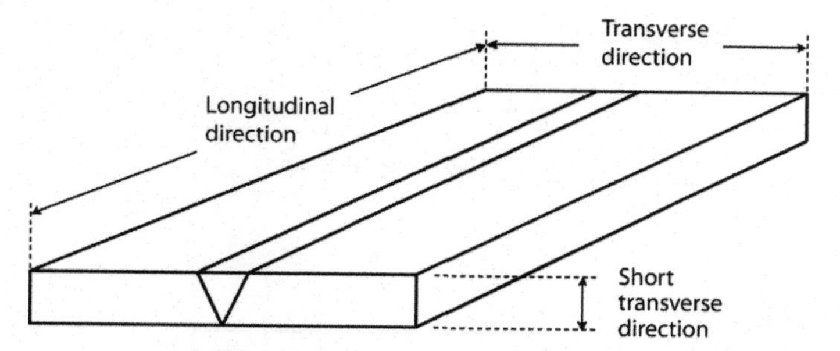

Figure 18.1 Main stress directions in a weld.

Residual stress in a weld occurs in three main directions (see Figure 18.1):

- Longitudinal
- Transverse
- Short transverse

Residual stresses in the weld in the longitudinal direction of the weld often amount to the yield strength of the material, or to just below it. In the transverse direction, the stresses in the weld are more dependent on the clamping condition of the parts. The transverse residual weld stresses are often relatively small.

18.3 Distortion

The uneven process of heating and cooling during welding also introduces potential distortions into the metal as it first expands and then contracts. If the adjacent parts cannot withstand this shrinkage, distortion will result, usually combined with residual stresses. There are several types of distortion:

- Longitudinal distortion (shrinkage perpendicular to the weld line)
- Transverse distortion (shrinkage parallel to the weld line)
- Rotational or angular distortion (rotation around the weld line)

Longitudinal and transverse distortions 'shorten' the weld (see Figures 18.2 and 18.4). An example of this kind of distortion is a welded beam that can become bent if the weld is not located symmetrically (in the centre of gravity of the cross section). If more than one weld is used, they must also be symmetrical.

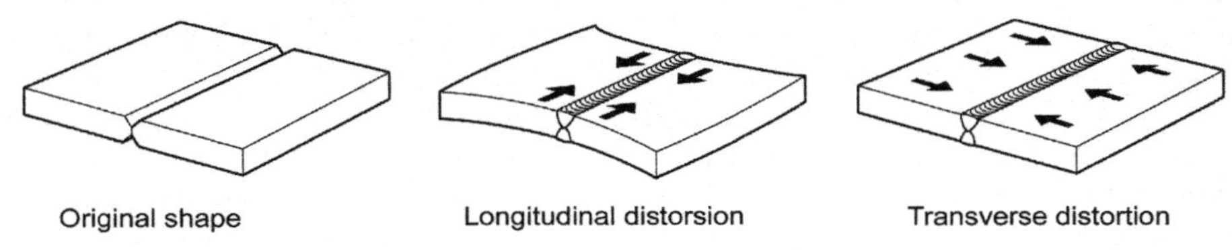

Figure 18.2 Longitudinal distortion and transverse deformation appears after welding.

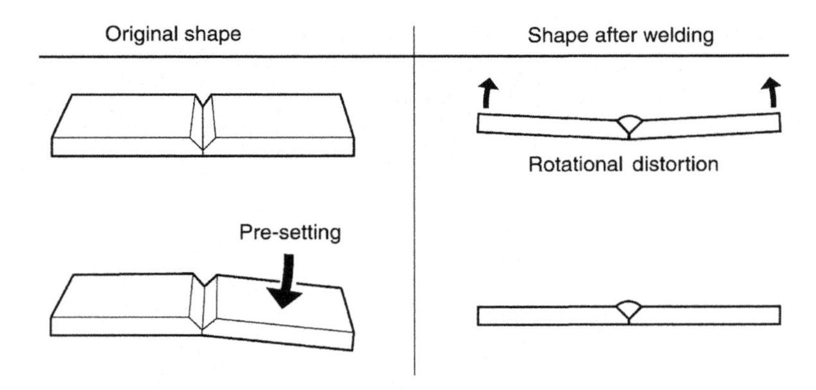

Figure 18.3 Rotational distortion can be prevented by pre-setting to compensate for distortion.

Rotational distortion bends the weld (see Figure 18.3). This kind of distortion can be minimised by making the weld bead symmetrical about the neutral axis or having a parallel-sided single pass weld, as with electron beam welding. A stiff section can also prevent this kind of distortion from appearing.

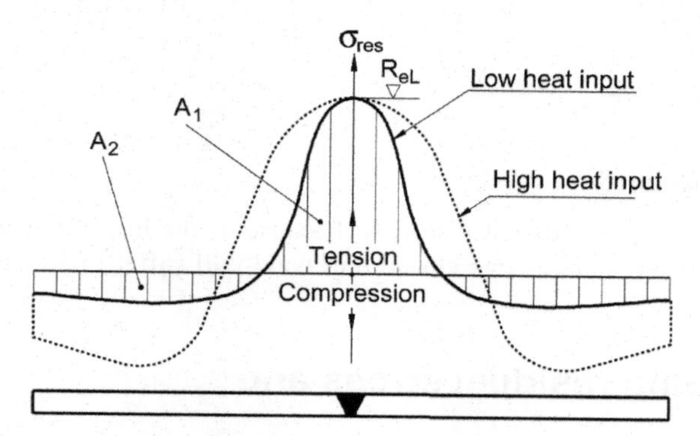

Figure 18.4 Longitudinal residual stresses in a butt weld. Area A1 = A2.

The problem of distortion is clearly affected by the characteristic of the weld metal. Figure 18.5 shows some key material properties that affect residual stresses/distortions. High thermal conductivity means that a larger area will be heated up by welding, thus increasing distortions. If the material has a high thermal capacity, more heat can be stored into it, also increasing the distortions. The greater the coefficient of thermal expansion of the material, the greater the distortions will be. Figure 18.5 shows differing levels of welding distortion for different materials. Stainless steels, for example, suffer higher levels of distortion than plain carbon steels.

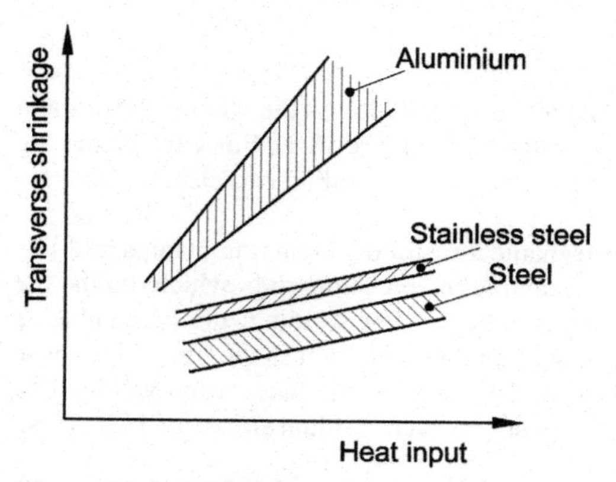

Figure 18.5 Schematic weld distortions.

TABLE 18.1 Some material data that affects the residual stresses/distortions of welds.

Property	Steel	Stainless steel	Aluminium
Coefficient of thermal conductivity (W/m°C)	approx. 50	approx. 25	235
Thermal capacity (J/kg°C)	approx. 450	approx. 500	920
Coefficient of thermal expansion (°C-1)	$11 \cdot 10^{-6}$	$17 \cdot 10^{-6}$	$24 \cdot 10^{-6}$

As well as the material being welded, the level of distortion is affected by such factors as:

- Weld preparation e.g. bevel angles
- Material thickness
- The amount of weld metal required

A single V butt weld will suffer significant distortion because of the high level of weld metal shrinkage. In contrast, a square-edge closed butt weld will suffer relatively little distortion.

18.4 Reducing welding residual stress and distortion

There are various ways of counteracting residual stress and distortion. These include:

- Appropriate design
- Presetting/offsetting
- Mechanical restraints
- Preheating
- Limiting the heat input and control of other weld parameters
- Thermal tensioning and heat sink welding
- Sequential welding
- Flame straightening/heat treatment
- Mechanical stress relief

In presetting/offsetting, pieces are offset to a preset angle. As the weld solidifies and shrinks, it pulls the pieces into the correct position (see Figure 18.3). Since the pieces are not restrained, residual stress is also reduced. This simple method can also be used, for example, on fillet welded T-joints.

By holding components in position, mechanical restraints including clamps and jigs can reduce the level of distortion. An example is precambering which holds the component in a predefined position during welding. After welding, the precamber is released and the component springs back to a minimally distorted shape. However, these techniques increase the level of residual stress. Preheating the parts to be welded can reduce residual stresses. If the parts are clamped, however, residual stresses will increase if the parts are preheated.

Higher heat input means that the zone in which residual weld stresses are high will be larger. In addition, compressive stresses a little way away from the weld will increase. Limiting the heat input can also reduce distortion. A more intense heat source allows higher speed, lower heat input and less distortion (see Figure 18.6). Fusion welding techniques often lead to the largest distortions while laser, electron beam and friction stir welding can result in lower distortions. Other weld parameters such as travel speed, single-pass versus multiple-pass welds can also influence stresses and distortions.

Thermal tensioning involves moving a heat source ahead of and/or alongside or behind the weld torch. This method can control distortions and residual stresses by controlling the heating and cooling rates of the weld. Heat sink welding is similar to thermal tensioning except a cooling source is used. An example of this method is pipe welding where it produces compressive weld residual stresses in the weld heat-affected zone along the inner surface of the pipe as a way of mitigating stress corrosion cracking.

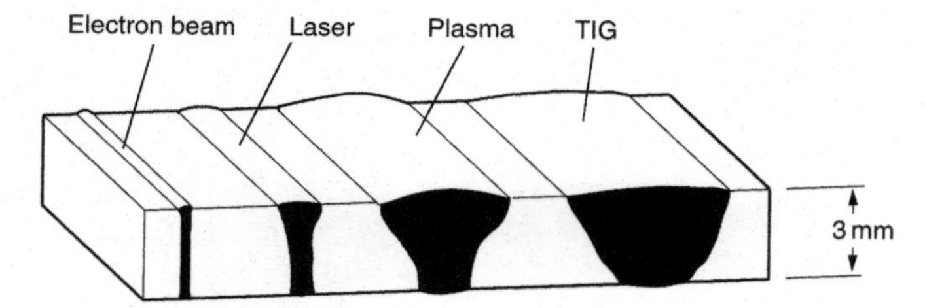

Figure 18.6 Penetration profile for some different welding methods.

Sequential welding involves welding in a way that allows parts of the workpiece to move during welding. Sequential techniques include balanced welding, back step welding and back skip welding.

Flame straightening uses a heating torch to provide an intense localised heat which straightens out distortions. Post-weld heat treatment (e.g. using heated jackets placed round components) can also provide stress relief, removing a high percentage of residual stresses. Mechanical stress relief techniques include the use of ultrasound or peening to redistribute residual stresses. Laser shock peening uses a high-density short-pulse laser to deform the surface and subsurface of a component, creating compressive residual stresses which can improve the damage tolerance of the component.

18.5 References and further reading

P. Michaleris (ed.), *Minimisation of welding residual stress and distortion,* Woodhead Publishing Limited, 2011.

Z. Feng (ed.), *Processes and mechanisms of welding residual stress and distortion,* Woodhead Publishing Limited, 2005.

19 The weldability of steel

19.1 Introduction

The commonest materials that are welded are steels, i.e. alloys of iron and carbon (Fe-C) as a base, with additives of a range of alloying elements. The main classes of steels discussed are:

- Carbon steels (plain carbon steel, carbon-manganese steel and micro-alloyed steel)
- High-strength steels
- Stainless steels (austenitic, ferritic, martensitic and ferritic-austenitic (duplex) stainless steels)

Weldability describes how far the materials affected by the weld retain the same strength, corrosion resistance, oxidation resistance etc. as the base material after welding. The following sections briefly describe the weldability of various steels.
The properties of the weld metal are determined largely by the choice of filler material, the type of base material, the welding method and the welding methodology, while the properties of the HAZ are determined primarily by the composition of the base material and the amount of thermal energy delivered during welding. Figure 19.1 shows the names of the various parts of the material affected by the weld.

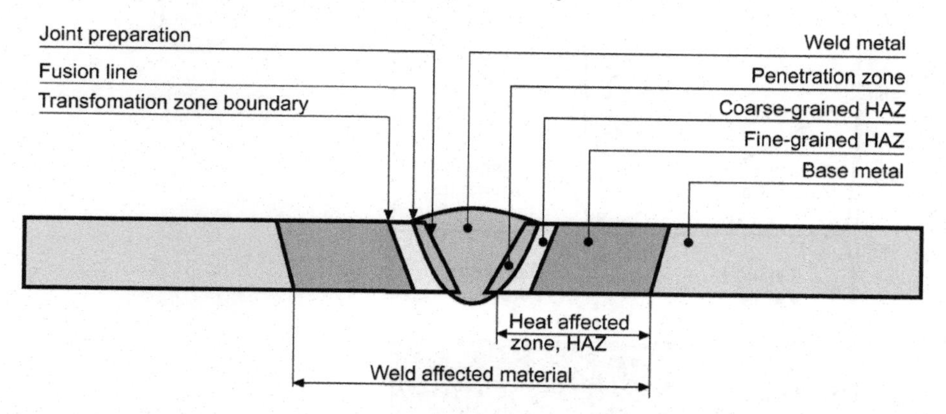

Figure 19.1 Nomenclature of zones and boundaries in heat affected zone.

19.2 Carbon steels

The International Institute of Welding (IIW) divides steels up into five quality classes, A, B, C, D and E, depending on the method of manufacture and impact strength (see Table 19.1). The European standard for general-purpose structural steel, EN 10025, gives a different classification for the steels, depending on their impact strength at various (low) temperatures.

As used here, the term 'carbon steels' refers to plain carbon steel, carbon-manganese steel and micro-alloyed steel. Plain carbon steels are characterised by alloying with up to 0.20–0.25 % of carbon to provide increased strength. They have largely been specially developed as structural materials for welded structures.

Carbon-manganese steels contain, in addition, up to about 1.5 % of manganese, with a maximum carbon content of 0.25 %. Examples of such steels include S235JR and S275JR according to the European steel standard EN 10025-2.

TABLE 19.1 Quality classes for plain carbon steels (Source: IIW).

Quality class	A	B	C	D	E
Degree of killing	Unkilled Semi-killed Killed	Semi-killed Killed	Semi-killed Killed	Semi-killed Killed	Killed
Impact toughness requirement, min. 27 J at:	–	–	0 °C	-20 °C	-40 °C

Microalloyed steel consists of a plain carbon steel or a carbon-manganese steel as the basic material, to which a small quantity (0.001–0.1 %) of one or more alloying substances has been added in order to reduce the grain size of the material and thus improve the yield strength and give better impact toughness. Examples of such grain-size-reducing elements include aluminium, niobium, titanium, zirconium and vanadium.

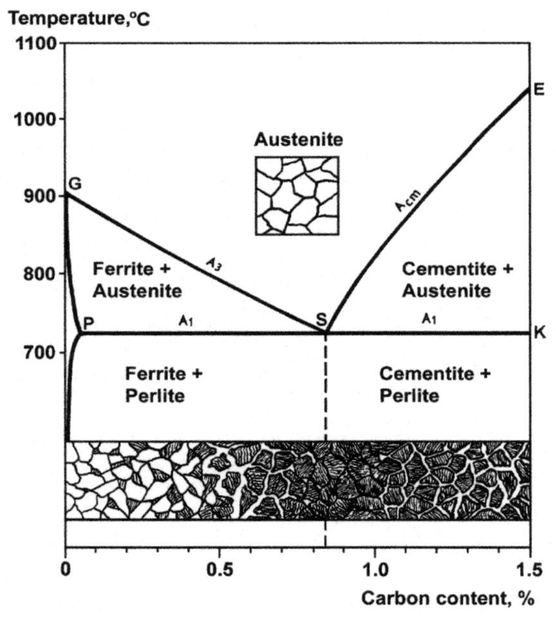

Figure 19.2 Part of the iron-carbon diagram, with a schematic representation of the structure.

How welding affects the structure of metal

The iron-carbon diagram (Figure 19.2) can be used to explain the origins of the various zones in those parts of the base material affected by welding. It can be seen from this that steels with low carbon contents are predominantly ferritic, while perlite (an eutectic mixture of ferrite and cementite) increases with increasing carbon content and cementite starts to appear in the grain boundaries at carbon contents over about 0.8 %. Bearing in mind what was said above concerning the composition of plain carbon steels, it can be seen that they will generally have a ferritic-perlitic structure. In the diagram, the PSK line represents the A_1 conversion temperature, the GS line represents the A_3 conversion temperature and the SE line represents the A_{cm} conversion temperature.

Figure 19.3 shows the primary structure of a single-pass weld. Heat flows quickly from the weld to the base material via the surface represented by the line abc, while the surface adc, exposed to the atmosphere on the top of the weld, loses heat less quickly. As the metal solidifies, the primary grains grow perpendicularly from the surface that is losing heat the most quickly, i.e. the melt face.

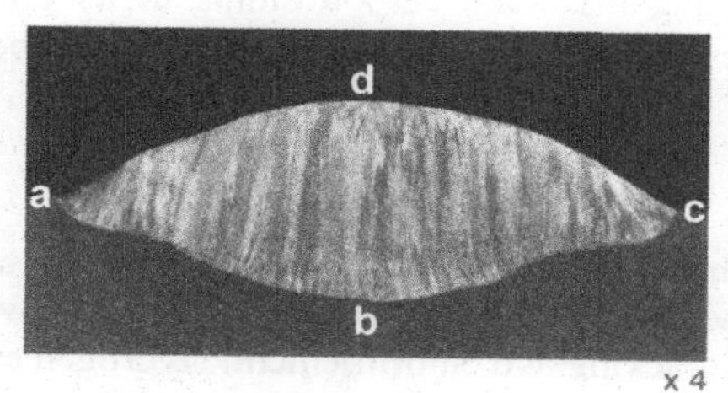

Figure 19.3 The primary structure can be seen in a section through the weld, x4.

Figure 19.4 shows schematically how the heat of the weld affects the base material closest to the weld. The structure of the rolled, unaffected base material is shown at the right, adjacent to the structural change zone, which has been heated to temperatures between A_1 and A_3.

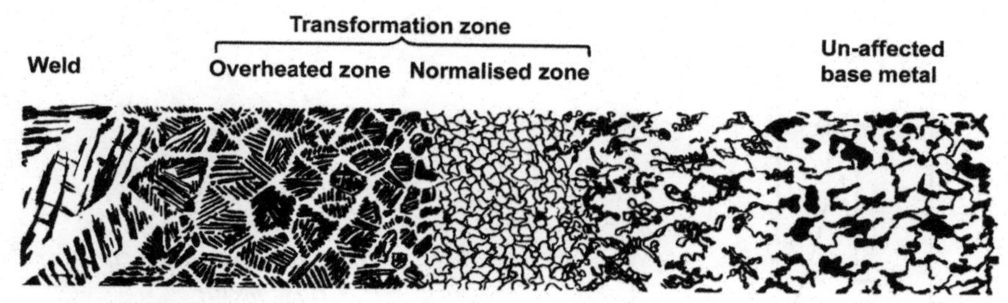

Figure 19.4 Schematic illustration of how the weld heat affects the base metal.

The transformation zone – often also referred to as the Heat Affected Zone (HAZ) – can be clearly seen adjacent to the weld. It can be divided into the normalised zone, which has been heated to somewhat over A_3, and the overheated zone, where temperatures have been considerably over A_3, up to the melting point of the material. The structure of the overheated zone exhibits a substantial amount of grain enlargement, and is partly Widmanstätten-orientated.

The weld metal can be seen on the left of Figure 19.4, in this case with a Widmanstätten structure, which is a characteristic of single-pass welds. In a multi-pass weld, each weld bead will be heat-treated by the subsequent bead. This means that, with the exception of the top pass, which retains its Widmanstätten structure, a multi-layer weld has what is known as a normalised weld structure.

When using a filler material, the properties of the weld metal are determined largely by the composition of the filler material. In the case of deep penetration into the base material, this also affects the properties of the weld metal. When welding without a filler, it is the suitability of the base material for welding that determines the properties and quality of the weld. Together with the heat input, it is also the composition of the base material that determines the changes in the properties of the base material in the HAZ. This applies in particular to the overheated zone which, in several respects, is the most critical area.

If the austenitic structure is cooled too quickly, ferrite and pearlite do not have time to form, allowing the formation of other microstructures with higher hardness but poorer ductility. If the material cools very quickly, martensite will be formed which can be described in simple terms as ferrite containing dissolved carbon atoms in its grain structure. Martensite is very hard and brittle, with the hardness increasing as the carbon content increases.

Risks of cracking

If the heat input is low in comparison with the thickness of the surrounding metal, cooling will be rapid and martensite may be formed. This means that hardness will increase and there may be a risk of hydrogen cracking and embrittlement. Hydrogen can be released into the weld metal from damp flux or damp electrodes, from damp or rusty metal surfaces, or from organic substances such as oil, paint or dirt. It diffuses into the surrounding base material from the weld pool, causing cracks when the material cools and its solubility declines.

It is generally accepted that hardness should not exceed 350 HV for carbon or carbon manganese steels though it can be higher for micro-alloyed steels. Figure 19.5 shows characteristic hardness contours around a weld.

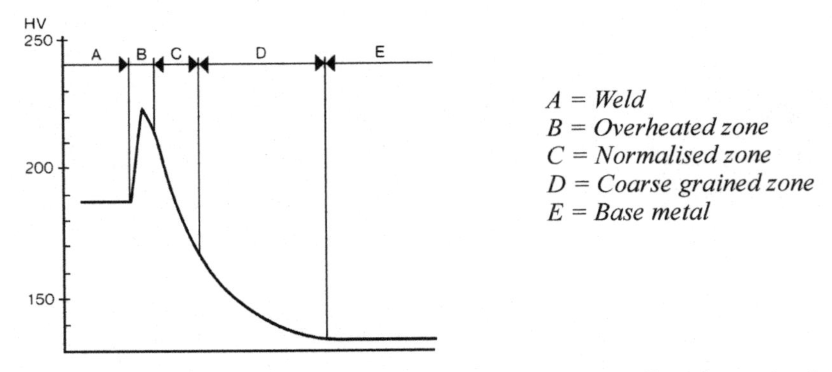

Figure 19.5 Typical hardnesses of various parts of the heat-affected zone in plain carbon steel.

To some extent, the risk of complications can be determined by calculation of the equivalent carbon content, *CE*, of the material:

$$CE = \mathrm{C} + \frac{\mathrm{Mn}}{6} + \frac{\mathrm{Cr} + \mathrm{Mo} + \mathrm{V}}{5} + \frac{\mathrm{Cu} + \mathrm{Ni}}{15}$$

A steel is regarded as fully weldable if *CE* does not exceed 0.41. At higher values, it is regarded as being weldable to a limited extent, which in general means that welding is performed at a certain temperature on the workpiece, in order to reduce the rate of cooling (see the standard EN 1011-2). Even if *CE* does not exceed these values, heating of the workpiece may still sometimes be necessary. For steels in the minimum strength group, this means that preheating is needed for metal thicknesses over about 50 mm, while for those in the highest strength groups it is needed for thicknesses over about 30 mm. In the case of plain carbon steels, preheating temperatures over 200 °C are unlikely to be needed.

As plain carbon steels have been developed for use in welded structures, there are not normally any problems if welding is carried out correctly. However, for 'non-standard' steels, and particularly for unkilled steels, there is a risk of thermal cracking if the material contains higher levels of contaminating elements. In addition, if a material not having a guaranteed impact toughness is used where such toughness is required, there is a risk of fractures due to brittleness in the HAZ.

Brittle failure can occur when stresses remain in the material after welding, in combination with stress concentrations around, for example, a weld defect, or if the temperature is below the transition temperature of the material. Thermal stress relieving reduces the crack sensitivity. Welded plain carbon steel structures can be stress-relieved at 550–600 °C when this is regarded as necessary.

19.3 High-strength steels

Steels having yield strengths up to 700 MPa, but meeting the $CE < 0.41$ weldability criterion, can be rolled in thicknesses from 2 mm to 10 mm. Most of these steels can also be cold formed, due to their low sulphur and carbon content. Cold forming allows parts to be produced that can then be joined to produce more complex structures, thus reducing the total amount of welding needed. Table 19.2 shows typical weldable cold-finished steels of this type.

Similar steels are also used for welded tubes and pipes, e.g. for bulk gas or oil mains, or for district heating distribution pipes. It can be seen from Table 19.2 that there are many strength classes of Domex steels. Hot-rolled steel of this type is also used for RHS sections. There is a European standard, EN 10149-2, for the hot-rolled flat products.

Extra high yield strengths, in combination with good weldability, can also be obtained by hardening and tempering or by accelerated cooling after rolling. In Sweden, high strength steels with acceptable weldability are manufactured by SSAB under the name of WELDOX. However, with yield strengths above 650 MPa, the *CE* value will exceed 0.41.

The geometry of a weld also affects the dissipation of heat. In the case of, for example, a fillet weld, where three plates meet, it is necessary to consider the combined metal thickness. Current rules for welding temperatures (EN 1011-2 method A) express the working temperature as a function of the combined thickness of the materials, the heat input, the hydrogen content of the weld metal and the carbon equivalent. Only filler materials with **guaranteed low hydrogen content** should be used.

Preheat temperatures may alternatively be calculated according to the carbon equivalent for the CET-method (EN 1011-2 method B) that is specially developed for high strength steels as WELDOX. At this method the joint geometry is based on just the thickest plate when all the plates in the joint are of the same steel. The CET-method is judged to be better than the CE-method for WELDOX steels.

TABLE 19.2 Steel grades from Domnarvet.

Grade	Yield strength R_{eH} MPa min	Ultimate tensile strength, R_m MPa min.	Micro-alloying elements
Domex 315 MC	315	390-510	Al + Nb
Domex 355 MC	355	430-550	Al + Nb
Domex 420 MC	420	480-620	Al + Nb
Domex 460 MC	460	520-670	Al + Nb
Domex 500 MC	500	550-700	Al + Nb + V(Ti)
Domex 550 MC	550	600-760	Al + Nb + V(Ti)
Domex 600 MC	600	650-820	Al + Nb + V(Ti)
Domex 650 MC	650	700-880	Al + Nb + Ti(V)
Domex 700 MC	700	750-950	Al + Nb + Ti(V)

TABLE 19.3 Steel grades from SSAB.

Grade	Thick-ness (mm)	Yield strength, *R*eH MPa min.	Ultimate tensile strength, *R*m (MPa)	Corresponding standards	Heat treat-ment
WELDOX 700	4-50 50-100 100-130	700 650 630	780-930 780-930 710-900	EN 10025-6	QT
WELDOX 900	4-50 50-80	900 830	940-1100 880-1100	EN 10025-6	QT
WELDOX 960	4-50	960	980-1150	EN 10025-6	QT
WELDOX 1030	4-12	1030	1200-1500		
WELDOX 1100	4-25	1100	1250-1550		
WELDOX 1300	4-10	1300	1400-1700		

WELDOX steel is available with yield strengths up to 1300 MPa, as shown in Table 19.3. Steels corresponding to the micro-alloyed hot-rolled Domex steels and the hardened and tempered WELDOX steels are also manufactured in Germany, the UK and France. However, the Swedish weldable high-strength steels have long been the leading steels of their type in the European market.

European standards for high-strength hot-rolled steels are being drafted in yield strength steps of 500 MPa. Work on an EN standard covering such aspects as lowest recommended working temperatures is also in progress.

19.4 Stainless steels

This group of steels consists of those that include chromium as an alloying element in concentrations of about 12 % or more. At these concentrations, the surface of the metal forms a thin, tough film of oxide, protecting the material from further corrosion or oxidation. The same mechanism increases the scaling temperature of the material, as compared with that of plain carbon steel. Other alloying elements are added in order further to improve corrosion resistance to various solutions: the most important of these elements are nickel and molybdenum.

Non-alloyed steels have a ferritic microstructure at room temperature. The relatively high concentrations of alloying elements in stainless steel have a significant effect on the microstructure. Alloying elements can be divided into two main groups, depending on their effect on the structure. They are referred to as austenite or ferrite forming elements. Austenite forming elements include nickel, manganese, carbon and nitrogen, while ferrite forming elements include chromium, molybdenum, silicon and niobium. The austenite forming elements extend the steel's austenite zone as concentrations increase, while ferrite forming elements reduce the zone as their concentrations increase.

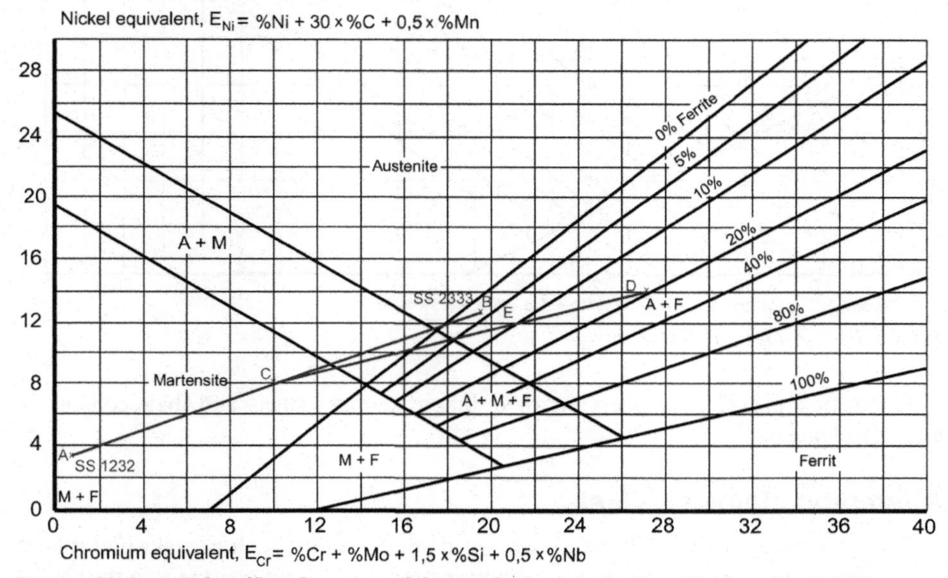

Figure 19.6 A Schaeffler diagram of the worked example described on Page 206.

The Schaeffler diagram shows how the structure of the material is affected by varying concentrations of austenite and ferrite forming elements. A variation of this is the DeLong diagram (Figure 19.7), which also shows the nitrogen content of the material. By calculating the equivalent chromium and nickel concentrations of the material from the diagrams, it is possible to estimate the structure of the weld metal. The diagrams do not apply for welds that have been heat-treated, nor for the base material affected by the weld.

Depending on the type and quantities of alloying elements, the structures of the materials in the group of stainless steels differ, giving them differing physical and mechanical properties, corrosion resistances and weldabilities. Stainless steels have been divided up into the following groups:

- austenitic stainless steels
- ferritic stainless steels
- martensitic stainless steels
- ferritic-austenitic stainless steels
- martensitic-austenitic stainless steels

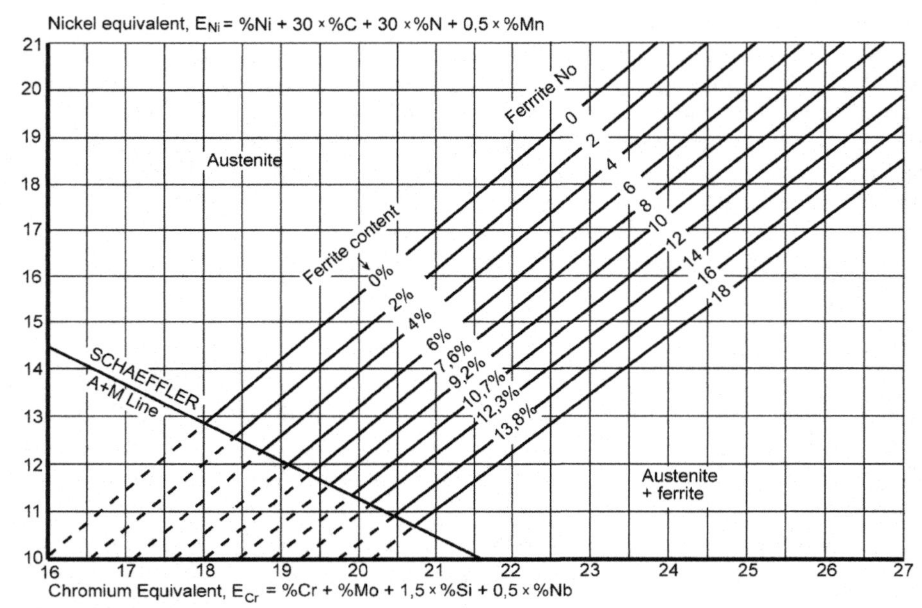

Figure 19.7 A DeLong diagram.

Table 19.4 provides an overview of ordinary stainless steels and their content of alloying elements.

Austenitic stainless steels

The main applications for these steels are for domestic items, such as sinks and saucepans, and for pipes and pressure vessels requiring good corrosion resistance in the chemical, cellulose, building and food industries. Austenitic steels contain about 12–27 % chromium and 7–30 % nickel. They are non-magnetic, and cannot be hardened by heat treatment. They are normally delivered in the solution-annealed state, which has required rapid cooling from a temperature of about 1050–1100 °C. Depending on the intended application, they often include about 2–3 % molybdenum as an alloying element, which increases corrosion resistance.

Carbon concentrations are normally below 0.05 %, or below 0.03 % for ELC steels (Extra Low Carbon). Certain materials have higher carbon concentrations, of about 0.08 %. In order to prevent chromium carbides from forming during welding, alloying elements in such steels must have a higher affinity for carbon than does chromium: suitable elements are titanium and niobium. Steels with this composition are referred to as stabilised steels. Figure 19.8 shows the microstructure of an austenitic steel after solution-annealing.

Although the heating and cooling caused by welding does not produce any significant structural changes in austenitic material, secondary phases can be formed under certain conditions in the weld metal and HAZ. The most important of these phases is ferrite, which has both beneficial and adverse effects on the properties of the weld. Figure 19.9 shows an austenitic weld with about 8 % ferrite.

TABLE 19.4 The main types of stainless steels (from MNC Handbook no. 4).

Group	Composition limits as given in Swedish Standards for steels within the group [1]				Harden-ability	Ferro-magnetic
	C %	Cr %	Ni %	Mo %		
Ferritic	0.08 0.08 0.20	12-13.5 16-19 24-28	- - -	 2.5	Cannot be hardened	Magnetic
Martensitic	0.09-0.35 0.14-0.23	11-14 15.5-17.5	– 2.5	1.2 –	Hardenable	Magnetic
Martensitic-austenitic	0.10 0.05	12-14 15-17	5-6 4-6	– 0.8-1.5	Hardenable	Magnetic
Ferritic-austenitic	0.030 0.10	18-26 24-27	4-8 4.5-7	0.1-5.0 1.3-1.8	Cannot be hardened	Magnetic
Austenitic – not including Mo as alloying constituent	0.12 0.07 0.05 stab.[2] 0.030	16-19 17-19 17-19 17-19 17-19	6.5-9.5 8-11 8-11 8-11 9-12 8-12	- - - - - -	Cannot be hardened	Non-magnetic
Austenitic – with Mo as alloying constituent	0.10 0.05 stab.[3] 0,030 0.025 0.020	16-19 16.5-18.5 16.5-18.5 16.5-18.5 19-28 19.5-20.5	8-10 10.5-14 10.5-14 11-17 24-34 17.5-18.5	1.3-1.8 2.0-3.0 2.0-2.5 2.0-4.0 3.0-5.0 6.0-6.5		
Austenitic, heat resistant	0.08 0.04-0.06	24.0-26.0 18.0-19.0	19.0-22.0 9.0-10.0	- -		

1)May include other deliberately added alloying elements, e.g. N, Cu, Al or Ce
2)Stabilised with Ti or Nb
3)Stabilised with Ti

Figure 19.8 The microstructure of an austenitic steel containing 18 % chromium and 8 % nickel. Magnification: 500 x.

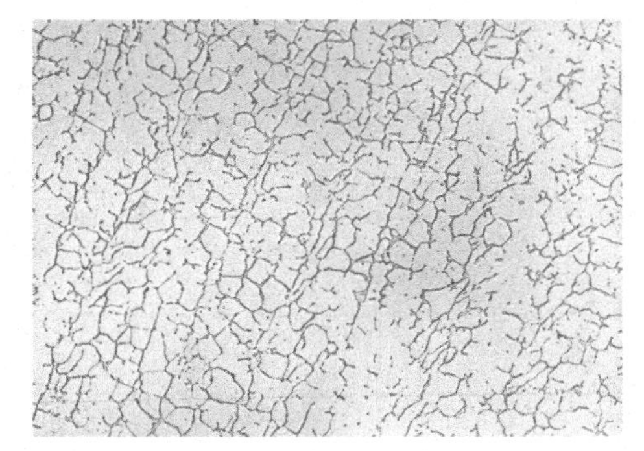

Figure 19.9 Section through an austenitic weld containing about 8 % ferrite: magnification, 500 x.

The beneficial effect of ferrite is that it largely prevents the formation of hot cracks in the weld, partly because it dissolves elements such as sulphur and phosphorus, which would otherwise segregate and substantially increase the risk of cracking as the stresses in the weld increase. Modern methods of steelmaking have made it possible to produce filler materials that give ferrite-free welds with good hot cracking resistance. They contain relatively high proportions of manganese and low levels of sulphur and phosphorus. The adverse effect of ferrite in the weld metal is that it can be selectively attacked by certain corrosive media. As the grain boundary ferrite is contiguous at concentrations of about 10 % and above, failures caused by corrosion can occur in such cases. This means that it is necessary to use a filler material giving considerably lower ferrite concentrations for structures likely to encounter such corrosion conditions.

At operating temperatures in the 550–900 °C range, ferrite is converted to what is known as the sigma phase, which reduces the ductility of the material. However, this is generally of little practical importance, if the ferrite content of the weld metal is less than about 5 %. A radical way of avoiding adverse effects of ferrite in the weld metal is, of course, to anneal the material at about 1100 °C, thus dissolving the ferrite and stabilising the austenite. However, this is often not feasible for practical reasons.

Intergranular corrosion

A major problem when welding austenitic steels has been the need to avoid carbide precipitations at the grain boundaries. These carbides are chromium-rich, and therefore reduce the chromium concentration in the austenite adjacent to the grain boundaries, thus creating the conditions for intergranular corrosion, as shown in Figures 19.10 and 19.11. Chromium carbides are formed more rapidly in steels with higher carbon concentrations, as can be seen in Figure 19.12. If the heating time at a certain temperature is longer than indicated by the respective curve, carbide will be precipitated, a process known as sensitising. Austenitic steel with a low carbon concentration can generally withstand the effects of this unfavourable temperature range for a sufficiently long time to allow welding to be carried out without risk of carbide precipitation. A material containing 0.056 % carbon, for example, can be held at 700 °C for up to about seven minutes without intergranular corrosion occurring during subsequent testing. Nevertheless, when working in thick metals, welding should be carried out with a controlled heat input. ELC steel is therefore chosen for certain applications, as it can be heated for about ten hours without carbide precipitation: an alternative is the use of stabilised material.

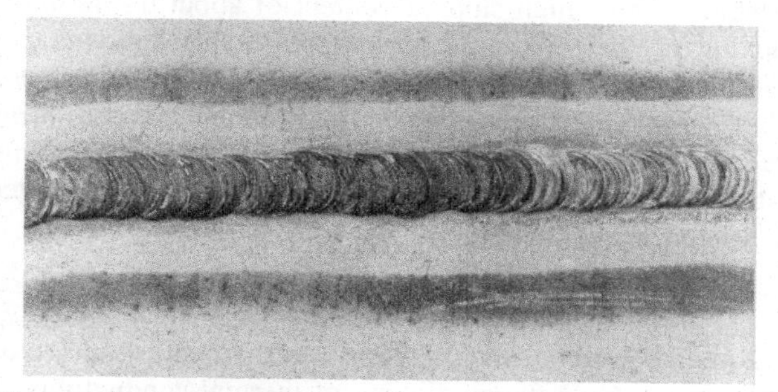

Figure 19.10 Intergranular corrosion at a weld in austenitic steel.

Stress corrosion

Under certain circumstances, austenitic steels are also sensitive to stress corrosion, e.g. caused by residual welding stress in materials in contact with chlorides or strongly alkaline media. Other materials should be used if it is felt that there could be a risk of stress corrosion: examples include ferritic-austenitic steels or austenitic steels with high nickel contents.

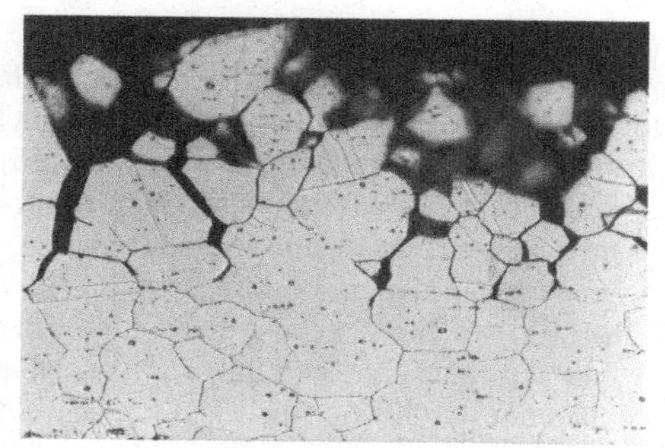

Figure 19.11 Section through an area of intergranular corrosion: magnification, 300 x.

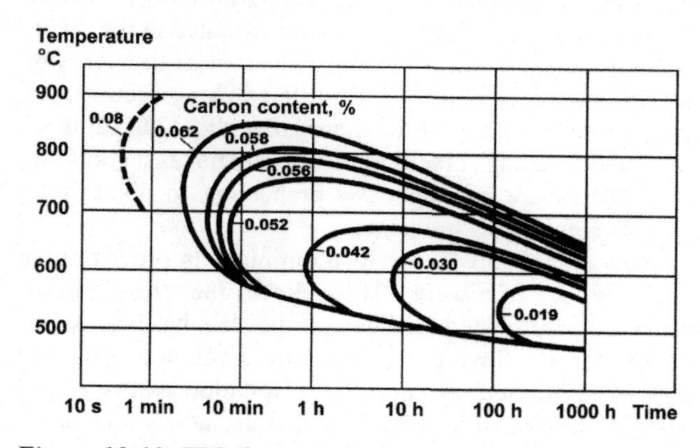

Figure 19.12 TTS diagram for austenitic steel with 18 % chromium, 8 % nickel and varying carbon concentrations. (TTS = Time-Temperature Sensitisation).

Tool steels

Austenitic tool steel, i.e. steel intended particularly for chip-cutting machining, is regarded as having poor weldability, as its high sulphur content of about 0.2 % introduces a risk of thermal cracking in the weld. However, tool steels containing a maximum of 0.05 % carbon, 0.20 % sulphur and more than about 1.5 % of manganese have relatively good weldability. As the manganese combines with the sulphur to form manganese sulphide, precipitation is avoided, which reduces the risk of cracking. This can be further guarded against by using fillers such as 18 8 Mn or 23 12 L. This latter filler metal has a ferrite concentration of about 15–20 %.

Welding stresses

In comparison with those of plain carbon steels, the coefficient of thermal expansion of austenitic steels is about 50 % greater, while the coefficient of thermal conductivity is about 40 % less. This means that there will be more contraction in a welded joint, so that welding must be carried out with particular consideration of possible distortion. It is generally sufficient to weld with a low heat input, with careful planning of the work and of the order in which welds are to be made, and to apply the welds symmetrically.

Stress relieving of the stresses introduced in the structure by welding is carried out at 850–950 °C or 400–500 °C. It is important that temperature ranges at which carbide can form are avoided in all heat treatment.

Ferritic stainless steels

Ferritic steels are used in many of the same applications as austenitic steels, i.e. in the food, chemical and cellulose industries. However, there are some applications in which their performance is superior, i.e. for use in sulphurous atmospheres, in which the presence of nickel reduces the resistance to corrosion, and in environments with modest chloride concentrations, where ferritic steels have better resistance to stress corrosion.

The main alloying element in ferritic steel is chromium, in proportions between 12 % and 30 %. Carbon content is generally less than 0.1 %. There are also Extra Low Interstitial (ELI) steels, with very low carbon and nitrogen concentrations, of less than 0.03 %, having better corrosion performance than normal ferritic steels and better weldability. Ferritic steels are magnetic and non-hardenable.

In comparison with austenitic steels, ferritic steels are not as easily welded. There is significant grain enlargement in the HAZ, with carbides being precipitated at the ferrite grain boundaries, resulting in embrittlement of the weld and reduction of its corrosion resistance. This embrittlement introduces a certain risk of brittle cracking occurring as the weld cools to the brittle transition temperature of the material at about 100–200 °C. In addition, there is a risk of transverse shrinkage cracks in the weld metal as it cools and stresses rise. These potential problems are countered by preheating to about 200–400 °C, which reduces the temperature and stress gradients.

Materials must be stress-relieved at 750–850 °C for 30–60 minutes in order to give optimum strength and corrosion resistance of a weld. This relieves the stresses introduced by welding and spheroidises the precipitated carbides. If the steel has been stabilised by the inclusion of titanium as an alloying element, stress relief to improve corrosion resistance can sometimes be omitted, as carbide precipitation that degrades corrosion resistance does not always occur when welding such metals; at any rate, not if the metal thickness is less than about 2 mm.

Steels containing more than about 20 % chromium form significant sigma phase ferrite at temperatures between 550 °C and 800 °C: if the steel contains molybdenum, this sigma phase formation extends to higher temperatures. This phase is brittle, and so substantially reduces the ductility of the metal and increases its ultimate tensile strength. Annealing in this temperature range is therefore inappropriate for these steels. The time needed to separate the sigma phase at any given temperature is very dependent on the composition of the steel, but sigma phase ferrite can form in weld zones when welding thick workpieces.

Another type of embrittlement, known as 475 °C embrittlement, can occur if the material is heat-treated in the temperature range of 400–550 °C.

Welding can be carried out, using a filler of the same material or of austenitic steel. The latter, with a nickel content of about 20 %, produces the best ductility of the weld.

Martensitic steels

Martensitic steels have chromium concentrations between about 12 % and 18 %. They contain more carbon than ferritic steels (about 0.1–0.3 %), which means that they can be hardened. This is done from about 1000 °C by quenching in oil or air, followed by annealing at 300–750 °C. These steels are magnetic.

Typical applications for them are where there is a need to combine hardness with corrosion resistance: examples include turbine blades, cutting tools, knives, razor blades, scissors, cleavers etc. Table knives and surgical instruments are other examples.

As these steels are air-hardening, there is always a hard and often brittle zone in the base material close to the fusion line, which means that these steels have poor weldability in comparison with that of other stainless steels. In order to reduce, to some extent, the hardness peak in this zone, these steels must always be welded at a prescribed preheating temperature. A suitable temperature depends on the material dimensions and type of joint/structure, but is generally in the range of 200–400 °C. In addition, welding must be followed by heat treatment at 650–750 °C for 30–60 minutes. If stress conditions in the structure are complicated, this heat treatment should be carried out immediately after welding, i.e. without allowing the workpiece to cool to room temperature between welding and heat treatment. However, in order to ensure complete metallurgical structure conversion of the metal, the temperature of the workpiece must not exceed about 150 °C when heat treatment starts.

If the workpiece is welded without preheating and then heat-treated, there is a risk of hardening cracks in the transformation zone. There is also a risk of brittle cracking, as the material has a relatively high brittle transition temperature, and welding-induced stresses in and around the weld zone are significant. In addition, there is a risk of hydrogen embrittlement in the HAZ, particularly if damp electrodes have been used. Further, if using a filler material of the same metal as the base material, there is a risk of transverse cracks in the weld metal, and so austenitic fillers are often used to reduce this risk. This also reduces the risk of brittle failure.

Particularly with complicated structures, it is usual to butter the joint faces with an austenitic filler metal. Each part of the workpiece is treated separately and heat-treated. The austenitic layer is built up to a sufficient thickness to ensure that no metallurgical change occurs in the base material when the final weld is made.

Ferritic-austenitic (duplex) steels

Ferritic-austenitic steels, which have attracted increasing interest since the 1970s, are often known as duplex steels. A characteristic of them is that they combine excellent corrosion resistance with high mechanical strength.

These steels find their main application area in the offshore industry, where their properties are particularly suitable for use in a high-chloride environment. Their high strength, in comparison with that of austenitic steels, also brings dividends in the form of lighter structures. The cellulose, chemical and petrochemical industries are other areas in which they are being increasingly widely used.

Steels in this group are characterised by chromium contents of up to 29 %, nickel contents of the order of 5–8 % and molybdenum contents around 1–4 %. The carbon content is usually low, at less than 0.03 %.

Another element that has shown itself to be very important in determining the weldability of ferritic-austenitic steels is nitrogen, which may be present in concentrations up to 0.4 %. When welding, the weld pool solidifies primarily with a fully ferritic structure, with the austenite forming at a later stage as the temperature falls. See Figure 19.13.

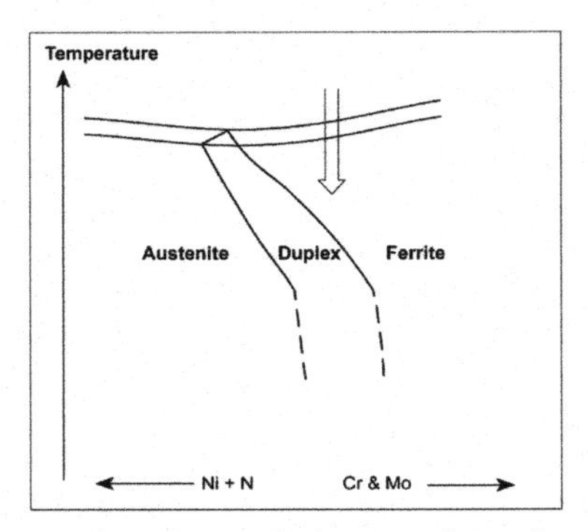

Figure 19.13 Pseudo-binary phase diagram for ferritic-austenitic steel.

The formation of austenite starts at the ferrite/ferrite grain boundaries, and grows inwards in the ferrite grains to produce a structure as shown in Figure 19.14.

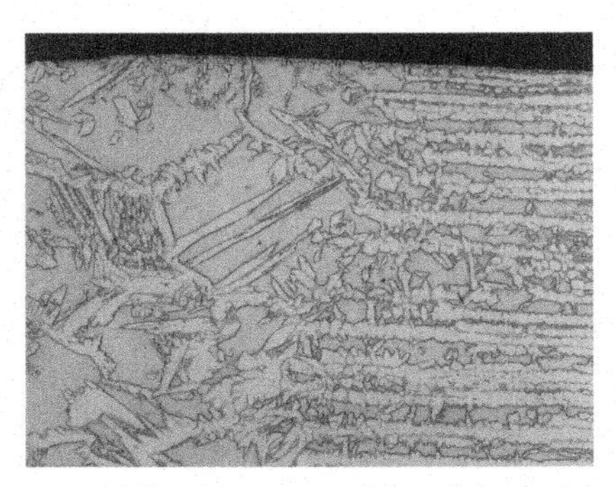

Figure 19.14 Microstructure of the weld metal and the HAZ in ferritic-austenitic steel.

The presence of nitrogen as an effective austenite forming element accelerates this process, to produce the correct metallurgical structure of the weld without requiring preheating or subsequent heat treatment. Older types of ferritic-austenitic steel, such AISI 329, with no nitrogen and a high proportion of ferrite, become almost entirely ferritic in a zone close to the fusion line, which degrades both the mechanical properties and corrosion resistance. It is necessary to provide preheating at about 100–200 °C to correct this. Ferritic-austenitic steels should not be welded at either too low or too high a heat input. A low heat input can result in too rapid a rate of cooling, which forms chromium nitrides that can adversely affect corrosion resistance. Too high a heat input, on the other hand, in combination with high temperature between passes, produces sigma phase (among other effects), which degrades both corrosion resistance and mechanical properties.

Welding should be performed using a filler of similar material, but with about 2–3 % higher nickel concentration in order further to assist austenite formation and to produce a ferrite concentration in the weld metal similar to that of the base metal. This ensures optimum characteristics of the weld metal in respect of corrosion resistance and mechanical strength.

Martensitic-austenitic steels

Applications for this steel group include water turbines, propellers, machine parts and structural steel in general.

These steels are magnetic, and have an approximate analysis of 0.04–0.08 % carbon, 13–16 % chromium, 5–6 % nickel and 1–2 % molybdenum. They can be hardened by quenching in oil or air from about 1000 °C, which produces a martensitic structure, possibly containing some residual austenite. Subsequent tempering at about 600 °C forms austenite, which is so stable that it is not converted to martensite as the material cools to room temperature. As a result, the final structure contains about 65–80 % of tempered martensite, with the rest being austenite, which means that the material has considerably better ductility than fully martensitic steel. These steels therefore have good weldability when tempered.

It is not generally necessary to preheat the workpiece, and nor is there any greater risk of cracking in the HAZ. That part of the HAZ that is heated to such a high temperature (over about 900 °C) that it is converted completely to austenite will be converted mainly to martensite by the subsequent cooling to room temperature. Tempering restores the martensitic/austenitic ratio of the base material in this zone.

The weldability of these steels is relatively good. Austenitic fillers are generally used, although this sometimes results in the strength of the weld metal being lower than that of the base material.

Welding of stainless steels to dissimilar metals

There is often a need, when fabricating stainless steel structures and/or when incorporating them in other plant, to weld stainless steel to plain carbon or low-alloy steels. In addition, different types of stainless steels often have to be welded to each other. In order to obtain the best performance of such joints, particular attention must be paid to the metallurgy of the weld metal. For this purpose, the Schaeffler diagram is a good help. In general, the following methods of working can be employed when deciding on the appropriate choice of filler material.

The position of the material in the Schaeffler diagram is determined from its chromium and nickel equivalents, and a straight line is then drawn between the points. Similarly, the position of the proposed filler material is plotted. When welding symmetrical joints, it can be assumed that the joint surfaces of the base material will be more or less equally melted. A straight line can therefore be drawn between the position of the filler material and the centre of the line between the base materials. In general, about 20–40 % of the filler material will be 'diluted' by the molten base material, with the result that the structure of the weld metal will be as indicated by a point about 20–40 % of the way along the line from the position of the filler material. If this structure is suitable, then the proposed filler material can be used. If it is not suitable, repeat the above procedure for a filler having a different composition.

The Schaeffler diagram in Figure 19.6 shows an example of the above for the case of a low-alloy steel EN 1.0401, (A), being welded to stainless steel 1.4301, (B), using 23 12 2 L filler (D). This gives a weld metal structure as shown at point E, i.e. austenite with about 10 % ferrite.

The filler material to be used for joining the two materials must also be selected as described above when welding compound materials consisting of plain carbon and high-alloy materials. The same applies for selection of the filler material for the first layer of stainless material to plain carbon or low-alloy steel.

19.5 References and further reading

J. Lancaster, *Metallurgy of welding*, Woodhead Publishing Limited, 1999.

G. Evans and N. Bailey, *Metallurgy of basic weld metal*, Woodhead Publishing Limited, 1997

N. Bailey, *Weldability of ferritic steels*, Woodhead Publishing Limited, 1994.

20 Welding of aluminium

20.1 Introduction

The difference between welding aluminium and welding steel is due to the different physical properties of the materials, as shown in Table 20.1.

TABLE 20.1 Comparison of the physical properties of aluminium and steel.

Property	Unit	Aluminium	Iron (steel)
Specific heat/weight	J/(kg·°C)	2	1
Specific heat/volume	J/(m^3·°C)	2/3	1
Heat of fusion/weight	J/kg	1.5	1
Heat of fusion/volume	J/m^3	0.5	1
Thermal conductivity	W/(m·°C)	3	1
Thermal expansion	1/°C	2	1
Modulus of elasticity	MPa	1/3	1
Density	kg/m^3	1/3	1

Special allowance must be made for the following properties of aluminium and its oxides:

- The substantial differences in thermal conductivity, thermal expansion and other properties of aluminium and steel mean that distortion occurs more easily in aluminium.
- Aluminium reacts strongly with oxygen, forming the oxide Al_2O_3.
- The oxide is strong and tough, and can easily cause weld defects.
- Due to its higher density, the oxide may sink into the weld pool, where it forms inclusions.
- The oxide does not melt during welding (melting temperature 2050 °C).
- Hydrogen dissolves readily in aluminium in the molten state, but not in the solid metal, and thus easily causes pores.

The oxide on the surface of aluminium therefore has a considerable impact on welding. If it is not removed before welding, it can cause weld defects, seriously reducing the strength of the weld.

20.2 Designation system for aluminium and filler materials

A standard system for classifying aluminium is that of the European Committee for Standardisation (CEN). Systems developed by CEN have the prefix EN. The first numeral in the numerical designation system developed by CEN indicates the main alloying element.

- 1xxx Unalloyed aluminium
- 2xxx Copper
- 3xxx Manganese
- 4xxx Silicon
- 5xxx Magnesium
- 6xxx Silicon + magnesium
- 7xxx Zinc
- 8xxx Miscellaneous

Wrought alloys (sheet, profiles, strip and foil) are indicated in this system by four numerals prefixed by the letters AW, while casting alloys use five numerals prefixed by AC. The systems are described in EN 573-1 (for sheet and profiles) and EN 1780-1 (for castings).

Group 1 is unalloyed aluminium of at least 99.00 % purity. The two last numerals in the group indicate the minimum aluminium percentage; e.g. AW-1070 indicates an aluminium alloy with 99.7 % aluminium and 0.3 % other elements, e.g. small amounts of iron and silicon. In the other groups, the second numeral indicates whether there has been any change to the original alloy. The last two numerals follow a numerical sequence.

The alloys in Groups 1, 3, 4 and 5 cannot be hardened by heat treatment, but those in Groups 2, 6 and 7 can be hardened by precipitation hardening (solution heat treatment and ageing).

For the various alloys, EN 515 specifies basic temper designations resulting from the materials' treatment:

F – as fabricated
O – annealed
H – strain-hardened, e.g. by cold rolling
W – solution heat-treated
T – thermally treated to produce stable tempers other than F, O or H

For the hardened materials in Group H, the standard provides further identification by means of numerals: examples are:

H12 – strain-hardened e.g. cold-rolled to quarter-hard state
H14 – strain-hardened e.g. cold-rolled to half-hard state
H18 – strain-hardened e.g. cold-rolled to fully hardened state
H24 – strain-hardened and partially annealed e.g. cold-rolled and heat-treated to the half-hard state

Additional designation numerals are provided also for Group T materials. Examples include:

T4 – solution heat-treated and naturally aged
T6 – solution heat-treated and then artificially aged
T7 – solution heat-treated and then artificially overaged

An example of a complete temper designation is EN AW-6063-T6, where EN stands for European standard, A for aluminium, W for wrought alloys (wire, sheet, profile). 6063 is an alloy in the 6000-group, consisting of aluminium alloyed with 0.4 % silicon and 0.7 % magnesium. Finally, T6 indicates that the material has been solution heat-treated and artificially aged (at elevated temperature).

An example of a cast alloy is EN AC-46200-F, where C stands for a cast material, and F indicates a casting not having received thermal treatment. See the relevant EN standards for further details of the designations. EN 573 describes the composition and form of the alloys: EN 515 defines the temper designations.

Classifications of filler materials

EN ISO 18273 gives classifications of filler materials for welding aluminium such as wire electrodes, wires and rods. It specifies the classification requirements for homogenous wire and rods for fusion welding of aluminium, basing the classification on the materials' chemical composition. Classification is divided into two parts:

a) the first part shows whether the product is homogenous wire or rods

b) the second part is a numerical symbol that indicates the chemical composition of the homogenous wire or rod.

The chemical composition limits of the filler materials are identical to those registered with the Aluminum Association in the USA for corresponding alloys. The symbol for homogeneous wire and rods is the letter S, e.g. S Al 4043. American standards specify that filler material specifications are preceded by the letters ER, e.g. ER 5183.

Typical fillers and their classifications are:

Al99.8 — Aluminium wire, very high purity. Example: S Al 1080.

Al99.7 — Aluminium wire of almost as high purity. Example: 1 S Al 1070.

Al99.5 — Aluminium wire of at least 99.5 % purity, for welding unalloyed aluminium of corresponding quality. Suitable for the chemical and foodstuffs industries.

Al99.5Ti — Aluminium wire of at least 99.5 % aluminium and with a few tenths of percent of titanium for high grain fineness in the weld. Example: S Al 1450.

Al99.0 — Aluminium wire of at least 99.0 % aluminium for welding unalloyed aluminium of corresponding quality. Suitable for the chemical and food industries.

AlSi5 — Silicon-alloyed welding wire with 5 % Si for TIG welding of aluminium containing not more than 7 % Si. Example: S Al 4043.

AlSi12 — Silicon-alloyed welding wire with 12 % Si for welding of aluminium containing more than 7 % Si. Example: S Al 4047.

AlMg3 — Magnesium-alloyed welding wire for welding aluminium containing not more than 3 % Mg. Example: S Al 5754.

AlMg4,5Mn — Magnesium-alloyed welding wire for AlMg alloys when high ultimate tensile strength is required. Resistant to salt water. Example: S Al 5183 (AlMg4,5Mn0,7).

AlMg4,5MnZr Magnesium-alloyed welding wire with a small amount of zirconium additive. For welding AlMg4,5Mn alloys. Examples: S Al 5087 (AlMg4,5MnZr) and S Al 5187 (AlMg4,5MnZr).

AlMg5 As AlMg4,5Mn, but with lower ultimate tensile strength. Non-cracking and resistant to salt water.

AlMg5Cr As AlMg5, but with a small addition of chromium for improved corrosion resistance. Example: S Al 5356.

AlMg5Mn Example: S Al 5556A.

AlMg5Mn1Ti Example: S Al 5556.

20.3 Weldability

Weldability refers to the suitability for the most common welding techniques. It is difficult to determine the weldability of aluminium alloys as weldability depends on the following (and other) factors:

- the type of alloy (i.e. the parent metal)
- the welding method
- the filler material
- the shape of the parts to be welded
- the degree of freedom of movement of the parts during welding.

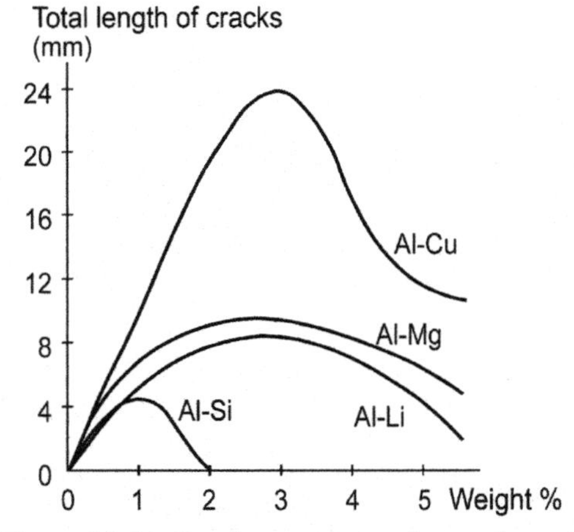

Figure 20.1 Crack sensitivity of a number of aluminium alloys when welding.

The sensitivity of the material to thermal cracking as the weld cools has been investigated for varying concentrations of a number of alloying substances (see Figure 20.1).

Aluminium materials are divided into those which are:

- easily weldable; materials that can be used in welded structures
- limited weldability; materials that should be avoided in welded structures
- unsuitable; materials that must not be used in welded structures.

The weldability of wrought alloys

Weldability refers here to the commonest material states. All alloys can be adhesively bonded, ultrasonically welded or joined by mechanical methods. The weldability of sheet and profile alloys is described in Table 20.2.

TABLE 20.2 Weldability of sheet and profile alloys.

	Range	Example of the alloy type	EN or AW designation
Good weldability			
– unalloyed Al	all types	Al99,7; Al99,5; Al99,0	1050A, 1200, 1350
– non-heat treatable alloys	most types	AlMn1; AlMg2,5; AlMg4,5Mn	3103, 5052, 5083
– heat treatable alloys	certain types	AlMgSi; AlSi1Mg; AlZn5Mg	6063, 6082, 7020
Limited weldability			
– heat treatable alloys	containing Cu and Pb	AlCuMg; AlCuMgSi; AlCuMgPb; AlCuZnMg	2011, 2014

TABLE 20.3 Non heat treatable alloys.

EN AW alloy	Gas welding	MMA welding	MIG/TIG welding	Resistance welding	Brazing	Soldering with flux
1060	A	A	A	B	A	A
1100	A	A	A	A	A	A
1350	A	A	A	B	A	A
3003	A	A	A	A	A	A
3004	B	A	A	A	B	B
5005	A	A	A	A	B	B
5050	A	A	A	A	B	B
5052, 5652	A	A	A	A	C	C
5083	C	C	A	A	X	X
5086	C	C	A	A	X	X
5154, 5254	B	B	A	A	X	X
5454	B	B	A	A	X	X
5456	C	C	A	A	X	X

A = Good weldability.
B = Acceptable weldability for most applications, but can require special methods or tests to find the correct welding parameters.
C = Limited weldability.
X = This method of joining is not recommended.

Alloys with a wide solidification temperature range are at greater risk of solidification cracking than those with a narrow solidification temperature range. The presence of copper or lead, and/or some other elements, as used in high-strength alloys and in those intended for lathe turning, also increases the risk of cracking. However, alloys containing more than 4 % copper can be welded. The weldability of heat and non-heat treatable alloys is described in Tables 20.3 and 20.4.

TABLE 20.4 Heat treatable alloys.

EN AW alloy	Gas welding	MMA welding	MIG/TIG welding	Resistance welding	Brazing	Soldering with flux
2014, 2017, 2024	X	C	C	B	X	C
2036	X	C	B	B	X	C
2090	X	X	B	B	X	C
2218	X	C	C	B	X	C
2219	X	C	A	B	X	C
2519	X	C	B	B	X	C
2618	X	C	C	B	X	C
6005, 6061	A	A	A	A	A	B
6009, 6010	C	C	B	B	X	C
6013	C	C	B	A	X	C
6070	C	C	B	B	X	C
6101, 6951	A	A	A	A	A	A
6262	C	C	B	A	B	B
6351	A	A	A	A	A	B
7004, 7005	X	X	A	A	B	B
7039	X	X	A	A	C	B
7075	X	X	C	B	X	C
7079, 7178	X	X	C	B	X	C

Weldability of casting alloys

The weldability of casting alloys can also be classified, particularly in relation to the type of filler material used (see Table 20.5). However, it is more difficult to weld cast items than it is to weld sheet or profile alloys, as castings often contain discontinuities in the form of gas inclusions, pores, shrinkage cavities etc, which can find their way into the weld metal as inclusions or pores, resulting in poor weld quality.

Castings can be welded to sheets or profiles, provided that there is good thermal balance between the parts. Gas or TIG welding should be used when making repair welds of faulty castings.

TABLE 20.5 Weldability and proposal of filler metal at welding of cast alloys.

	Cast alloy EN	Type of filler metal
Good weldability		
– AlSi	42000, 42100, 43100, 44100, 47000	AlSi5 AlSi12
Limited weldability (preheated to 150–250 °C)		
– AlSiCu	46200	AlSi5
– AlMgSi	51400	AlSi5; AlMg5
– AlZnMg	71000	AlMg5; AlMg4,5Mn
Not recommended (substantial risk of cracking)		
– AlMg (%Mg>6)		
–AlCu		
– Die castings in	44300, 44400, 46000, 46500	

20.4 Suitable methods of welding aluminium

Aluminium can be welded by gas welding and MMA (coated electrodes), but, due to the dependence on flux and the low thermal input, these methods are not widely used. They have been replaced by gas arc welding with inert gas shielding. MIG and TIG are the methods most commonly used currently, completely dominating the market for welding materials up to 50 mm thick.

Friction stir welding

Almost all aluminium alloys are weldable with FSW. Welding speed is affected by the strength of the material at elevated temperatures. FSW is a form of local heat treatment, in which the temperature of the material never reaches its melting point (see also Page 150). The weld has a fine grain structure, with typically equiaxed subgrains of 3-6 mm in size. The fine-grained microstructure of the weld/HAZ can be easily shaped.

Joint preparation

The commonest joint types for TIG or MIG welding of aluminium are shown in Figure 20.2. Apart from edge joints, welds are assumed to have full penetration. Backing should be used if the fit of the two pieces is poor, or when it is not possible to perform a last root pass. All joints are suitable for manual or mechanised welding.

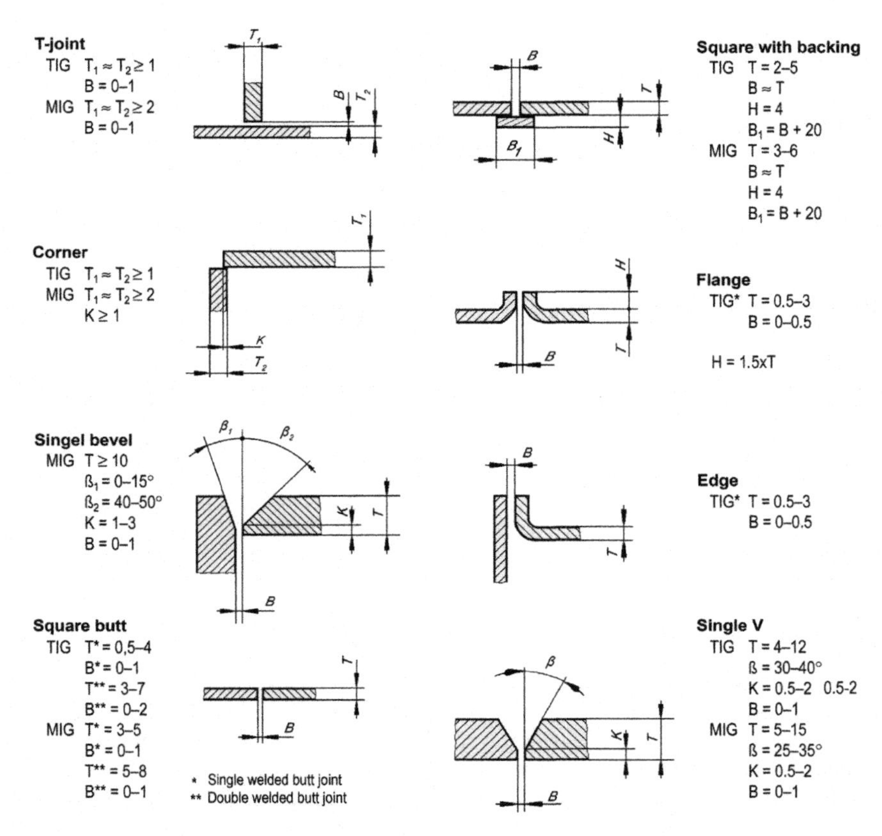

Figure 20.2 Joint types for MIG and TIG welding of aluminium.

20.5 Filler materials

Filler materials are used for two reasons: to provide additional material to fill the gap, and to supply alloying substances to avoid cracking as the weld cools. If the material has only a low risk of cracking, addition of alloying elements is not needed. Filler materials are therefore selected on the basis of both type of alloy and joint geometry. The correct choice of filler material can produce a welded joint with optimum properties in respect of strength, corrosion resistance and absence of cracking. MIG welding is always performed with filler wire, while TIG welding can be performed with or without filler wire (see Table 20.6).

Filler materials for plate and profile alloys

For AlMgSi alloys, AlMg5 gives the best strength, while AlSi5 has the lowest risk of crack formation and is easiest to weld. AlSi5 should not be used for AlMg alloys. It is important to note that

- the filler materials themselves are not heat treatable, and so the weld cannot be hardened by heat treating after welding.
- AlSi5 cannot be used where a good colour match is required between the weld and the parent metal after anodising, as the weld is dark grey in colour.
- Compromises have to be made when welding sheet and profile alloys to castings. AlSi5 or AlMg5 are often the best choice.

TABLE 20.6 Examples of filler materials for TIG and MIG welding of aluminium.

Type of filler material	EN	R_m parent metal MPa	R_m weld metal MPa	Hardness of the weld (HB)
Al99,8	1080A			
Al99,5	1050A	> 70	75	27–32
Al99,5Ti	1450	> 70	90	25–30
AlMn1	3103			
AlSi5	4043	>150	165	55–62
AlSi7Mg	4018			
AlSi10Mg	4046			
AlSi12	4047	>160	170	60–70
AlMg2Mn0,8Zr	5249			
AlMg3	5754			
AlMg3Mn	5554	>200	230	55–65
AlMg4,5Mn0,7	5183	>280	285	70–80
AlMg4,5MnZr	5087	>280	285	70–80
AlMg5	5019	>260	265	70–80
AlMg5Mn	5556A	>290	295	72–82
AlMg6MnZr			330	

Filler materials for casting alloys

The commonest casting alloys contain high concentrations of silicon. The basic rule for choice of filler material is then to use AlSi wire. AlSi12 should be used if the casting material contains more than 7 % of silicon, or AlSi5 if it contains less than 7 % of silicon. AlMg5 should be the first choice when welding magnesium alloy and AlZnMg castings.

20.6 Strength after welding

The structure of any aluminium material that has been affected by the heat of welding will always be altered (see Figure 20.3). The often rapid cooling changes the structure of the material away from that produced by normal heat treatment. The structure of the weld itself is that of a cast material, with the parent metal mixed with the filler material, while the areas closest to the weld will have been rapidly heated to the soft annealing temperature. The material will thus have been affected so that its strength and corrosion resistance properties will have been changed, often adversely. The area affected by the heat is referred to as the Heat-Affected Zone (HAZ).

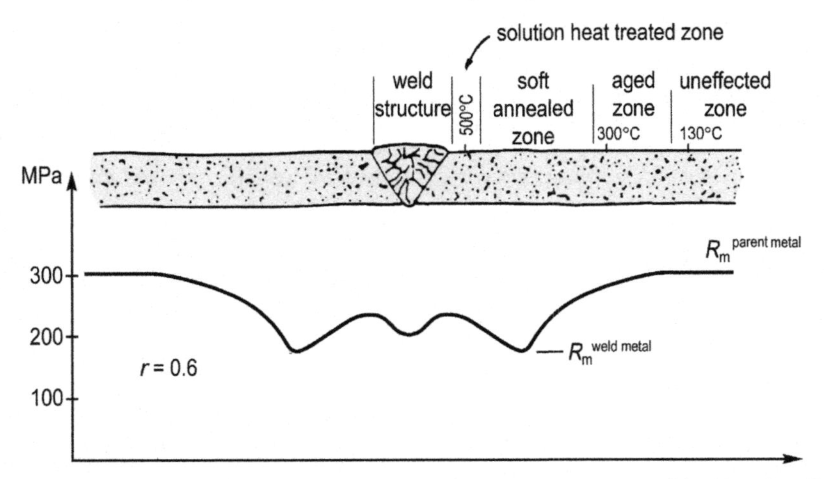

Figure 20.3 Strength distribution through the weld. Heat treatable AlMgSi alloy.

When welding annealed material, the strength of the material cannot be further reduced, but the strength of part-hardened materials (1/4-hard, half-hard etc.) or of aged materials (naturally or artificially aged) can be reduced after welding. Soft-annealed condition must be expected in the HAZ, and so joints should be positioned in areas of low loading. The ultimate tensile strength of a welded joint can be expressed by the reduction factor *r*.

$$r = \frac{R_m^{\text{weld metal}}}{R_m^{\text{parent metal}}}$$

The magnitude of the factor depends mainly on the condition of the material before welding, and also on the type of alloy, the method of welding and the speed of welding. Examples of pre- and post-welding strength. Table 20.7 gives values of *r* for MIG and TIG welding.

The loss of static strength can be partly recovered for certain alloys by the following procedures:

- cold working of the weld and the HAZ, e.g. by peening or rolling. This method is most suitable for non-heat treatable alloys, but can result in distortion of the weld zone.
- artificially ageing of AlMgSi alloys can give a 10-20 % increase of Rm.
- solution heat treated + artificially ageing restores 85-100% of the original strength, but causes distortion during cooling after the solution stage.
- cold ageing of AlZnMg alloys restores 85-90 % of the original strength.
- artificially ageing of AlZnMg alloys restores 85-100 % of the original strength, and also reduces stress corrosion.

However, in practice, these methods offer limited recovery of lost strength. Designers must therefore be aware of this, and position the welds where they will be exposed to a minimal load.

A practical way of obtaining the strongest possible weld is to exploit the opportunities offered by extrusions, to design profiles with integrated joint preparation and backing (see Figures 20.4 and 20.5). This reduces the number of welds and makes it possible to incorporate joint preparation in the profile.

TABLE 20.7 Examples of pre- and post-welding strength.

parent metal	EN AW-	Temper	$R_m^{\text{parent metal}}$ MPa	R_m^{weld} MPa	r
Non-heat treatable alloys					
Al99,0–Al99,7	1070A	O	70	70	1,0
	1200	H14	100	70	0.7
		H18	140	70	0.5
AlMg1	5005	H14, H24	130	100	0.75
		H18	160	100	0.6
AlMg2,5	5052	O	180	180	1.0
		H14, H24	230	180	0.8
		H18	260	180	0.7
AlMg4,5M	5083	O	280	280	1.0
		H12	300	280	0.95
Heat treatable alloys					
AlMgSi	6063, 6063A	T4	130	105	0.8[1]
		T6	215	130	0.6
AlSi1Mg	6082	T4	205	160	0.8[1]
		T6	290	190	0.7[2]
				(170)[3]	(0.6)[3]
AlZnMg1	7020	T4	320	280[4]	0.9
		T6	360		0.8

1. Artificially ageing after welding gives $r = 0.9$
2. Solution annealing and artificial ageing after welding give $r = 0.9$
3. Values in brackets ( ) are for TIG welding
4. After 90 days at room temperature or 60 hours at 60 °C or 3 days at room temperature + 24 hours at 120 °C

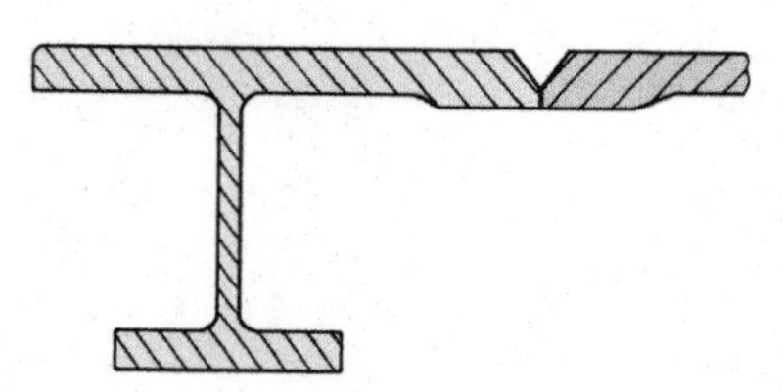

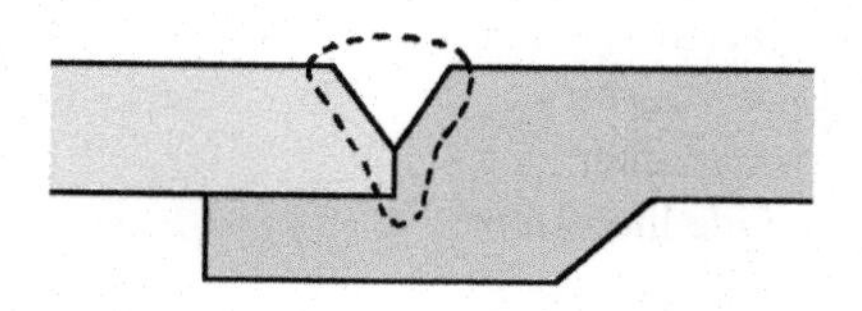

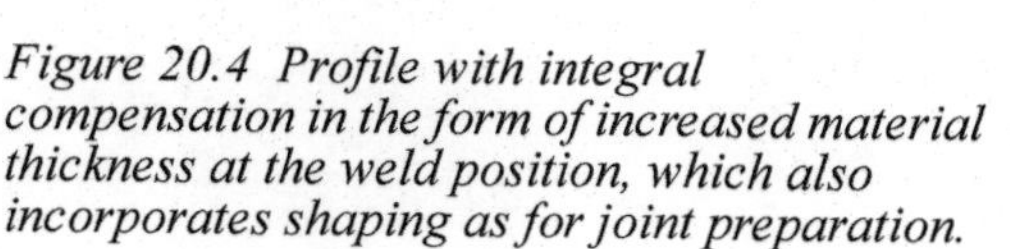

Figure 20.4 Profile with integral compensation in the form of increased material thickness at the weld position, which also incorporates shaping as for joint preparation.

Figure 20.5 Integrated backing in an extruded profile.

20.7 Quality issues in aluminium welding

Distortion

Welding an aluminium item usually causes greater deformation than that which occurs with steel. This requires modification of the welding method, which adds to cost. Good knowledge of deformation mechanisms therefore improves control of the design and reduces manufacturing costs.

In order to cause as little deformation as possible when welding, high-performance welding methods with low heat input should be used whenever possible. The rate of welding should be as high as possible, which means that MIG welding is preferable to TIG welding.

Corrosion resistance of welded material

The normally good corrosion resistance of aluminium materials is affected to some extent by welding. This is due partly to the heat of welding adversely affecting the corrosion resistance of the metal closest to the weld, and partly to the coarse-grained cast metal structure of the weld itself. Thermal cracks left unrepaired are particularly dangerous: corrosive media easily find their way into them, so that corrosion can continue in the interior of the cracks.

The corrosion resistance of pure aluminium and non-heat treatable alloys is only slightly affected by welding. The corrosion resistance of heat treatable AlMgSi-type alloys, which normally have the best corrosion resistance in the hot-aged state, on the other hand, is adversely affected by the heat of welding, and the same applies for AlZnMg alloys. However, heat treatment after welding restores the material to the best corrosion resistance state.

Discontinuities in welds

Discontinuities, form deviations, faults and defects are terms that are used in connection with weld quality. *Discontinuities* are cracks, slag inclusions, pores and poor fusion. *Form deviations* are undercuts, weld convexities and improper root reinforcement. Both discontinuities and geometric deviations can be *approved* or *not approved.*

The term 'weld defect' is often used as a general term. However, this should be avoided, as it can be misunderstood. Properly, a weld defect means that specified requirements have not been met. The commonest discontinuities arising when welding aluminium are:

- cracks
- poor fusion
- root defect
- oxide inclusions
- undercuts
- pores
- incomplete penetration
- shrinkage
- cavities

Discontinuities in aluminium welds are usually caused by inadequate joint preparation and cleaning. Incorrect welding parameters can also cause problems. *Cracks* and *poor fusion* are serious discontinuities, but the most common are *pores*. Cracks and poor fusion are not normally accepted, and must be repaired, but small numbers of pores can often be accepted. However, this depends on the required class of weld.

Cracks are not generally caused by the welding procedure, but more by the choice of parent and filler materials. Inadequate clamping, faulty jigs, poor tack welds and incorrect welding sequence can also cause cracks. It is not uncommon for x-ray examination to show that cracks have started from pores, shrinkages or oxide inclusions.

Ways of avoiding defects

There are number of ways of reducing defects. These include:

- Remove the oxide
- Ensure good gas shielding
- Inspect the feed rollers
- Adjust the welding gun angles

Aluminium is a reactive metal, which means that it quickly reacts with oxygen in the air, to form a thin layer of oxide, which provides good corrosion protection. With its high melting point of 2050 °C (which should be compared with the melting point of pure metallic aluminium at 658 °C), this oxide affects the welding process. Although the arc, in both DC MIG welding and AC TIG welding, breaks down the oxide layer, the presence of oxides still means that there is a risk of weld defects. The joint and nearby areas must therefore be carefully cleaned by means such as filing, scraping, brushing or washing. The edge of the material is generally overlooked when removing oxide, which increases the risk of poor fusion. Files and stainless steel wire brushes are the commonest tools used. When preparing an I-joint, the reverse of the plate must also be cleaned in order to prevent root defects.

God gas shielding is important for weld quality. It is not sufficient merely to ensure the correct gas quantity: the gas must also emerge uniformly from the gas cup. Gas spreaders and contact tips must be properly cleaned and defect-free.

The pressure of the feed rollers on the wire must not be too high. Wire feeders used for aluminium welding should have more than two drive rollers in order to avoid deformation of the soft aluminium wire. Aluminium should preferably be welded with a push/pull wire feeder system that both pushes and pulls the wire.

Having the correct forehand welding angle, 10-15°, of a MIG welding gun is very important. The wrong angle can result in inadequate gas shielding, or can even result in sucking air into the arc. Backhand welding generally produces convex weld strings.

Pre and post treatment

Cleaning the material prior to welding is important in order to achieve the best results. Dirt, oily marks, grease, moisture and oxides must be removed before welding starts, either by mechanical or by chemical means. Chemical cleaning may be necessary for higher standards of cleanliness. The following procedure can be used for ordinary workshop handling:

1. Removal of dirt and degreasing by alcohol or acetone

2. Mechanical removal of oxide by a) wire brushing (stainless steel rotating brush) b) scraping, or c) blasting

It is very important that, after welding, the weld and all parts that have been affected by the heat, spatter etc., should be carefully cleaned. This is important, as aluminium's normally good corrosion resistance will have been reduced by the welding.

After-treatment can be mechanical, e.g. blasting, wire-brushing (stainless steel wire brush), grinding or polishing. Do not use the same equipment on aluminium that has been used on steel.

Cleaning prior to surface treatment should consist of degreasing and/or pickling. Degreasing can be performed in hot alkaline aqueous solution, which etches the surface so a small amount of metal is removed.

Chemical surface treatment can be performed for the dual purposes of improving the resistance of the surface layer to attack and as pre-treatment prior to painting or other organic surface treatment.

Anodising aluminium is a common method of improving corrosion resistance, and can be combined with colouring of the surface.

Blackening

If the welding data is wrong, it can cause blackening of the weld and the metal in the vicinity. This black deposit consists of aluminium and magnesium oxides which deposit on the weld. Most of this blackening can be avoided by correct setting of the weld data.

20.8 References and further reading

G. Mathers, *The welding of aluminium and its alloys*, Woodhead Publishing Limited, 2002.

Swedish Welding Commission: *Goda råd vid aluminiumsvetsning.*

Author: Staffan Mattson

21 Design of welded components

21.1 Introduction

Determining the required design parameters of a welded joint can be relatively time-consuming, and sometimes complicated. This chapter is aimed at the design/production unit, with the objective of explaining how the designer thinks when designing a welded joint. Only exceptionally are absolute design rules referred to: instead, the presentation concentrates on practical advice and suggestions, aimed at producing an optimised design in terms of strength, weight, performance, ease of manufacture and overall cost. Those who are interested in more detailed design requirements should turn to references [1], [2], [3], [6] or [7]: see Page 242.

21.2 Symbolic representation of welds on drawings

A welding symbol on a drawing consists of:

- An arrow line (1)
- One or two reference lines (2)
- An elementary symbol (3)
- Possible supplementary symbols
- Dimensions of the weld

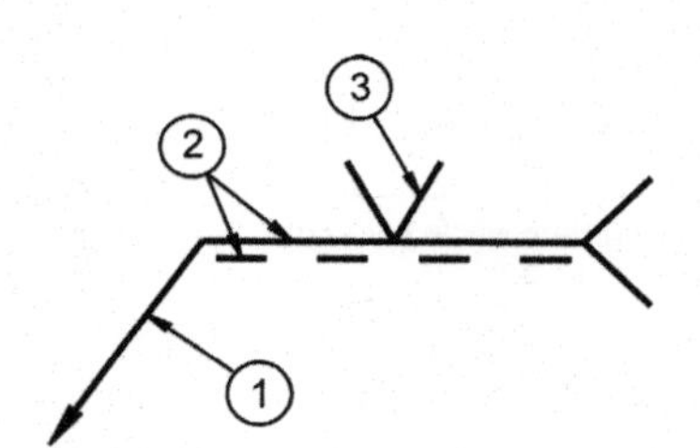

Figure 21.1 Symbols used on welding drawings.

Symbolic presentation of welds on drawings are given in ISO 2553, reference [4].

No	Designation	Weld	Symbol
1	Weld in plates with raised edges		
2	Square butt weld		||
3	Single V-butt weld		V
4	Singel-bevel butt weld		
5	Single-V butt weld with broad root face		Y
6	Single-bevel butt weld with broad root face		
7	Single-U butt weld (parallell or sloping sides)		
8	Single-J butt weld		
9	Backing run		
10	Fillet weld		
11	Plug weld		
12	Spot weld		○
13	Seam weld		

Figure 21.2 Examples of elementary symbols.

Elementary symbols and supplementary symbols

In general, the elementary symbol is similar in shape to that of the welded joint (i.e. before welding, indicating how the metal sheets are to be prepared for welding). Examples of elementary symbols are shown in Figure 21.2. If the unbevelled edge exceeds 2 mm, the joint is a single-V butt joint with broad root faces (Y). If not, it is a single-V butt joint.

Supplementary symbols may be used, in combination with the elementary symbols: see Figure 21.3. Absence of supplementary symbols means that there are no specific requirements in respect of the shape of the weld surface (except the requirements given in EN ISO 5817).

Shape of weld surface	Symbol
a) Flat (finished flush)	—
b) Convex	⌒
c) Concave	‿
d) Toes shall be blended smoothly	⅃L
e) Permanent backing strip	M
f) Removable backing strip	MR

Designation	Illustration	Symbol
Flat single-V butt weld		▽ with —
Convex double-V weld		X with ⌒
Concave fillet weld		◺ with ‿
Flat single-V butt weld with flat backing run		V with — above and below
Single-V butt weld with broad root face and backing run		Y with backing run
Flush finished single-V butt weld		√ ▽ 1)
Fillet weld with smooth blended face		◺ with ⅃L

1) Symbol in accordance with ISO 1302; instead of this symbol the main symbol √‾ can be used.

Figure 21.3 Supplementary symbols.

The importance of the reference lines

The position of the elementary symbol on the reference lines indicates on which side of the arrow line that the weld is to be placed. The upper, solid line (which is recommended to be terminated by a tail showing that the representation refers to ISO 2553) indicates a weld on the arrow side. In this case, the elementary symbol is placed on the solid line. The lower, interrupted line indicates a weld on the other side. In this case, the symbol 'hangs' below the interrupted line. See Figure 21.4 and Figure 21.5.

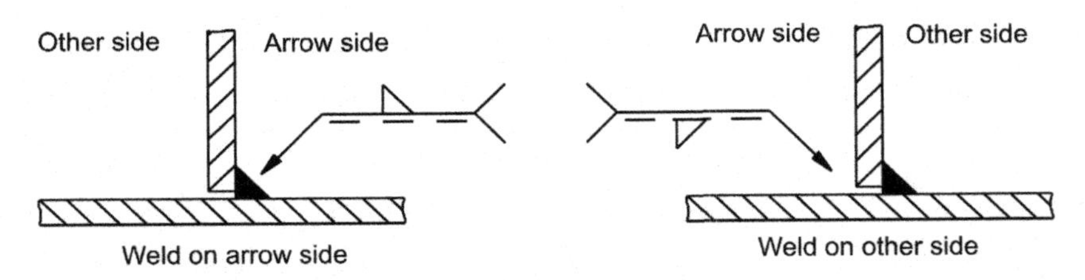

Figure 21.4 A T-joint with one fillet weld.

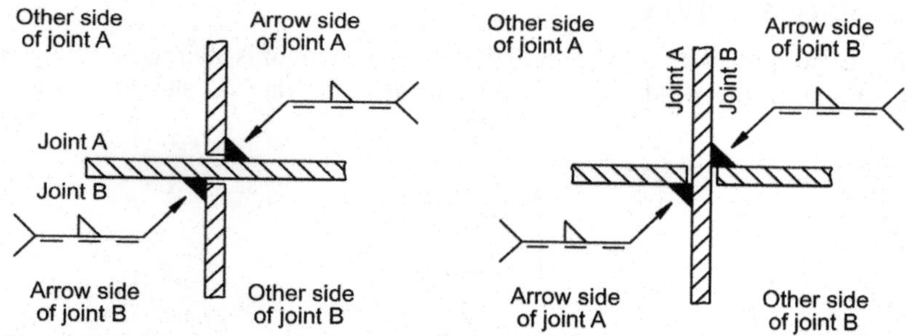

Figure 21.5 A cruciform joint with two fillet welds.

The interrupted reference line is not used for fully symmetrical welds: examples are shown in Figure 21.6.

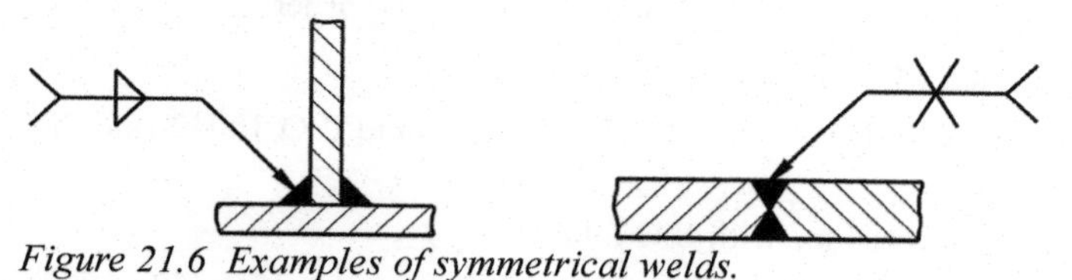

Figure 21.6 Examples of symmetrical welds.

The position of the arrow line

In general, there is no significance in the position of the arrow line in relation to the weld, except in the case of single bevel butt welds and single-J butt welds where the arrow of the arrow line must point towards the plate that is prepared. See Figure 21.7.

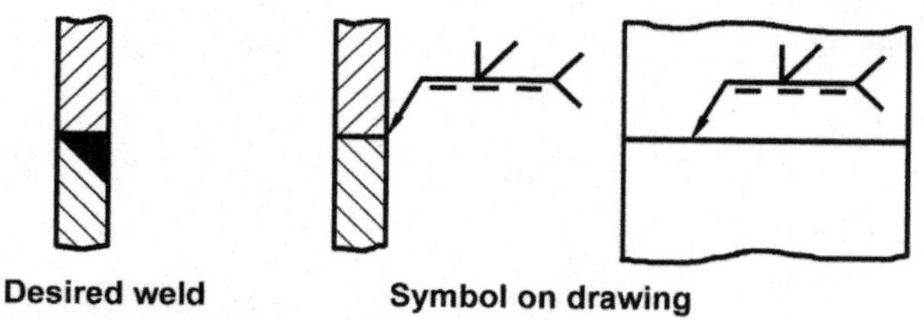

Figure 21.7 Position of the arrow line.

Dimensioning of welds

The dimensions of the cross-section of the weld are shown to the left of (before) the elementary symbol (e.g. penetration for butt welds, leg length or throat thickness of fillet welds). Write the length of the weld to the right of (after) the basic symbol.

E.g.: 5||300 indicates a square butt weld, with 5 mm penetration and a length of 300 mm.

z10◺ indicates a continuous fillet weld with a leg length of 10 mm.

z8◺ 5x200 (100) indicates an intermittent fillet weld with a leg length of 8 mm, divided up into five 200-mm-long welds, spaced 100 mm (free space between the welds) apart.

Complementary symbols

When a weld is to be applied all the way round a part, the symbol is a circle as shown in Figure 21.8a. Field or site welds are shown by means of a flag, as shown in Figure 21.8b.

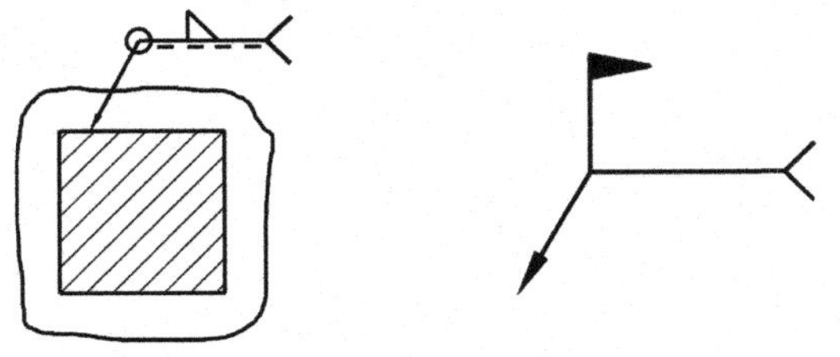

Figure 21.8 a) Peripheral weld. b) Field or site weld.

Further information can be given after the tail, in the following order:

- process (e.g. in accordance with ISO 4063);
- acceptance level (e.g. in accordance with ISO 5817 (steel) and ISO 10042 (aluminium));
- working position (e.g. in accordance with ISO 6947);
- filler metal (e.g. in accordance with ISO 544:2003, ISO 2560:2009, ISO 3581:2003).

The various items should be separated by slashes (/).

In addition, reference can be made to specific instructions (e.g. a procedure sheet) using a symbol in a closed tail, as shown in Figure 21.9.

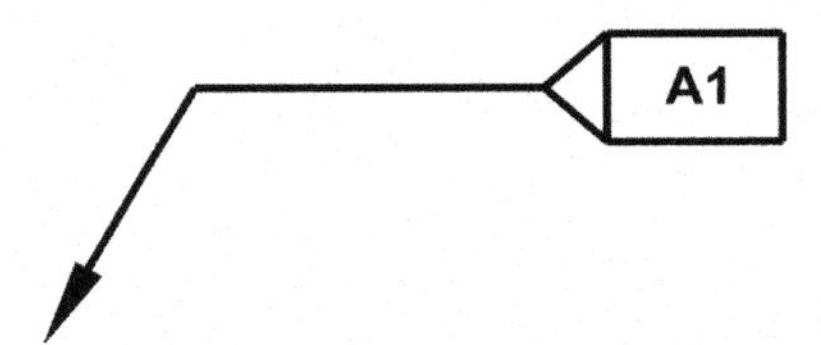

Figure 21.9 Reference information.

21.3 Welding classes

The welding classes are given in ISO 5817, reference [5].

If the designer has specified a welding class on a drawing, this has been done in order to ensure that the production unit is aware of the quality requirements applicable to the joint. ISO 5817 specifies three different quality levels: B, C and D, with Class B being the strictest. The standard includes tables that specify different types of defects, and how large they are permitted to be in the various quality levels.

Class D is usually used for non-load-carrying parts or in structures subjected to only low static loads.

Parts subjected to high static loads are normally welded according to Class C.

Parts subjected to fatigue loading are normally welded in Class B, with the additional requirement that the transition between the weld and the workpiece materials must be smooth. In extreme cases of fatigue loading, there may be a requirement that the edges of the weld must be ground or TIG-treated to produce a rounded shape (e.g. with a radius of at least 4 mm).

21.4 Design considerations

Load and stress distribution

When subjected to a load, a stress flow spread through, and acts on, the various parts of the structure. Changes in the geometry interfere with this stress flow, giving rise to stress concentrations, as shown in Figure 21.10 a–d.

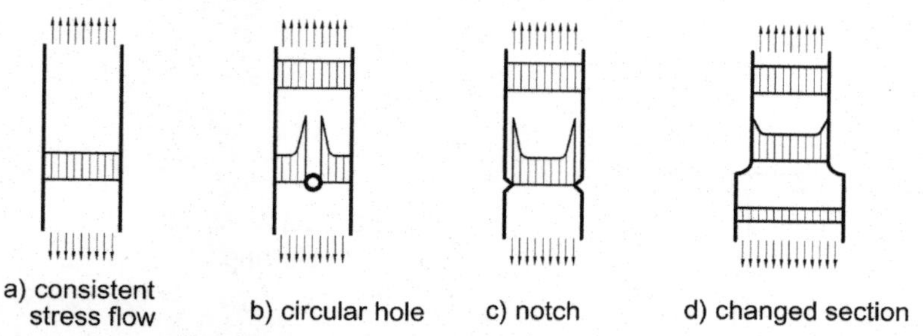

Figure 21.10 Examples of changes in stress flow.

These stress concentrations are of less importance in statically loaded structures. Although, strictly, some parts of the material may be stressed to beyond their yield strength limit, this does not actually involve any safety risk, as parts of the material will simply yield and redistribute the stresses.

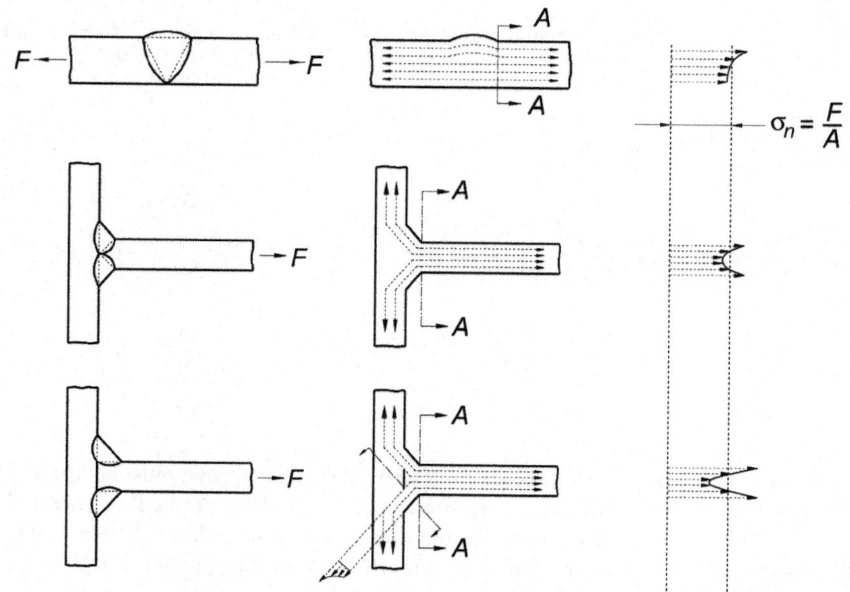

Figure 21.11 Schematic stress flow in various types of welded joints.

The situation is different, however, in structures subjected to fatigue loads. In such cases, the stress concentrations are vital in determining the overall strength of the structure, so that care must be taken in the design to avoid stress concentrations. As far as welded joints are concerned, the welds themselves constitute a stress concentration, as shown in Figure 21.11.

Design to transfer local forces

A main rule of design is that parts should be arranged so that forces are transferred in the plane of the material, and not perpendicular to it. If an applied force acts in the plane of the material, then the material is used to its maximum (resulting in tensional, compressive or shearing stresses).

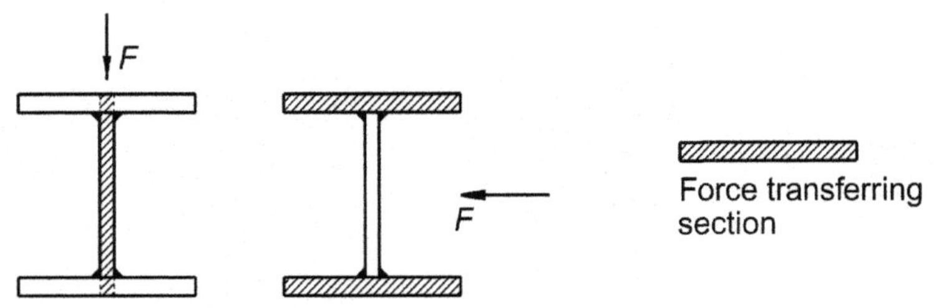

Figure 21.12 Marked surfaces act like shells.

If, instead, the load is applied perpendicularly to the plane of the sheet, it will act like a plate and be subjected to bending, which in turn means that all of the material cannot be used to its maximum. Figure 21.12 shows the parts transferring the forces in an I-beam.

If follows from this, that if a hanger eye is to be attached to a beam, it should be designed as shown in Figure 21.13.

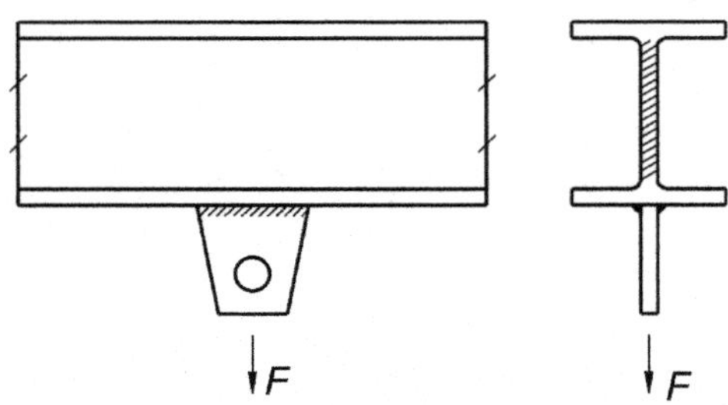

Figure 21.13 Beam with a hanger eye.

Figure 21.14 is an example of how not to design. If the hanger eye must be arranged perpendicular to the web of the beam, reinforcements could be applied as shown in Figure 21.15, which will transfer the load to the web of the beam. This will then replace the undesirable load on the bottom flange by a more favourable application of the force in the plane of the reinforcements.

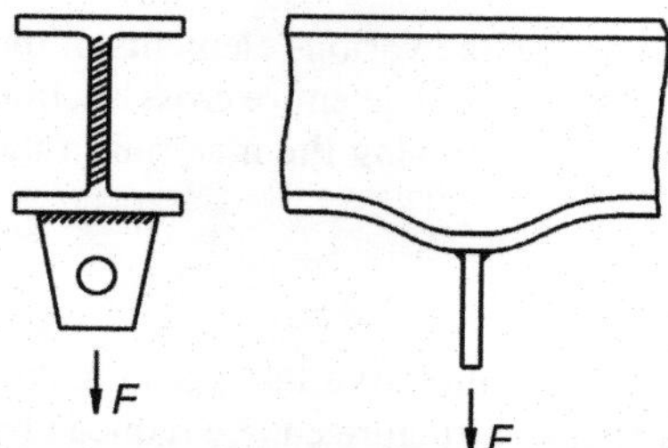

Figure 21.14 A hanger eye welded to the flange of a beam - poor design.

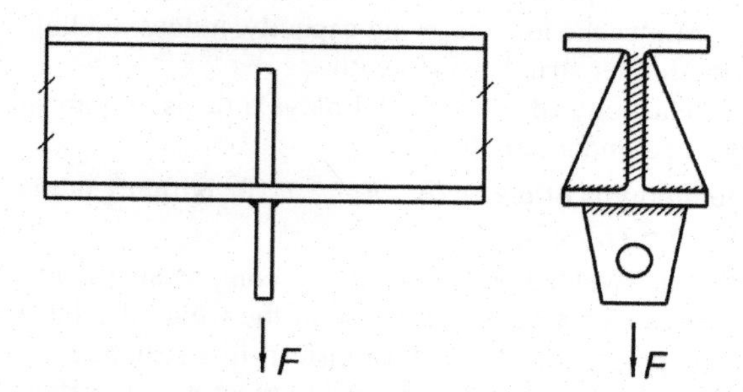

Figure 21.15 Reinforcements transfer the load to the web of the beam.

Figure 21.16a shows a completely wrong position for positioning a hanger eye, giving rise to high bending stresses in the web. Figure b shows a better way in which the forces can be transferred into the plane of the flanges.

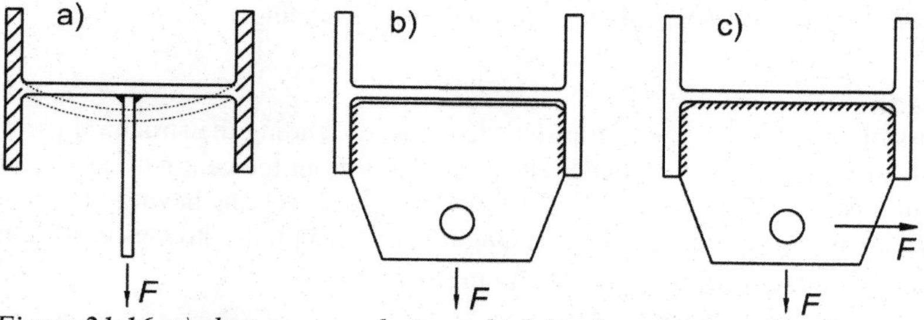

Figure 21.16 a) shows a poor design, which has been improved as shown in Figure b). If the beam is subjected to horizontal forces, the attachment should be arranged as shown in Figure c).

Design as determined by the type of load

The following order of priorities can be recommended for dealing with the various types of load encountered in a structure:

1. Tensile loads
2. Compressive loads
3. Shear loads
4. Bending loads
5. Torsional loads

Tensile loads

An excellent result will be achieved if it is possible to design the various elements of the structure such that they are subjected mainly to tensile stresses. The entire cross-section of the material then plays its part in carrying the load, thus utilising the material in the optimum manner to produce light, cheap designs.

Compressive loads

Arranging for the loads in the structure to be carried as compressive loads also makes good use of the material. However, the strength of a slender structure can be reduced by the risk of buckling or other instability phenomena. The critical buckling load is independent of the strength of the material. This means that, when designing slender structures in which the risk of buckling determines the load-carrying capacity, it does not help to choose an alternative material having higher structural strength.

On the other hand, the modulus of elasticity of the material plays a decisive part in determining the load-carrying capacity of slender structures.

All structural steels have the same modulus of elasticity, $E = 2{,}1 \cdot 10^5$ N/mm^2, while that of aluminium is much lower at $E = 0{,}7 \cdot 10^5$ N/mm^2.

In the case of slender structures that are welded and subjected to compressive loads, the longitudinal residual compressive stresses acting on each side of the weld, also have a negative effect on the load-carrying capacity. In the case of cistern-like structures in particular, the true buckling stress is considerably lower compared to theoretically calculated buckling stress (if consideration is not taken to residual stresses).

Shear loads

When part of a structure transfers the load by shearing, it acts as a shell, which is favourable: see "Design to transfer local forces" on page 226. However, there is a risk of shear buckling in the case of thin walled shells. The load-carrying capacity will then be reduced in a similar way as for buckling under compressive loading.

Bending loads

If it is not possible to avoid having to transfer the loads by bending the structural parts, the first step is to attempt to place the material as far away from the centre of gravity of the cross section as possible. Figure 21.17 shows two cross sections having the same cross-sectional areas. If exposed to equal bending moments (M_b), the stresses in section b) will be many times higher compared to those in section a).

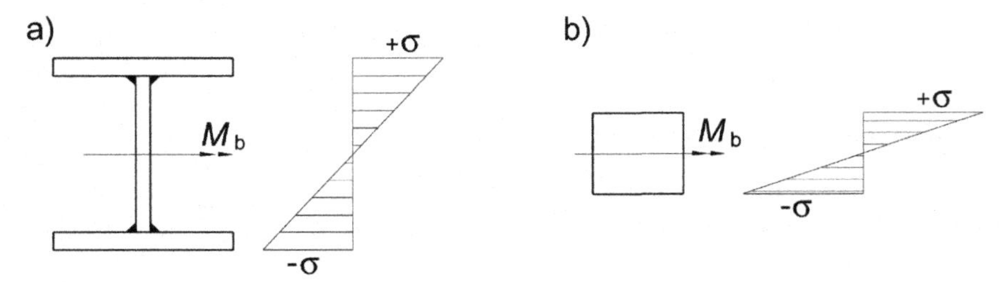

Figure 21.17 Bending stresses.

Torsional loads

It is particularly unfavourable to attempt to transfer torsional loads in open, thin-walled sections. Welded structures are often of this type. If, instead, the opening can be closed, torsional resistance will be substantially increased, as shown in Figure 21.18.

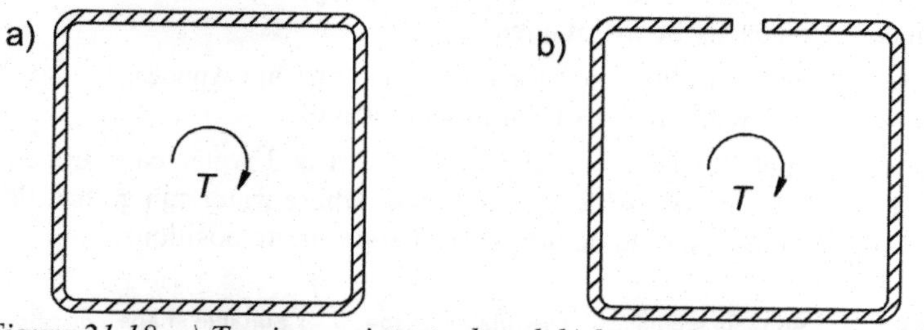

Figure 21.18 a) Torsion-resistant, closed b) low torsion-resistance, open.

Every cross section has a point – centre of torsion – to which an external force can be applied *without* imposing torsional loads upon the section. Figure 21.19 shows examples of the position of the centre of torsion in a number of common sections.

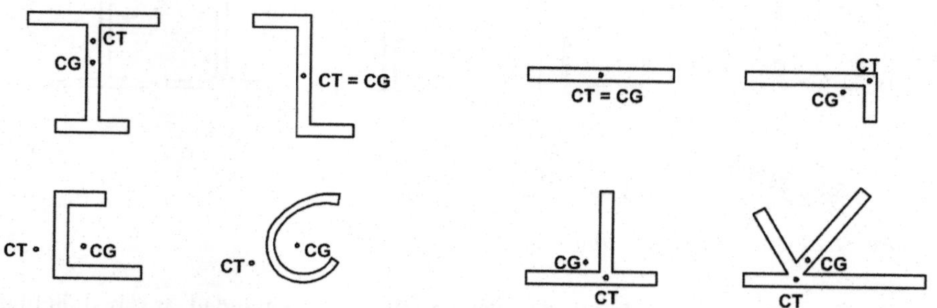

Figure 21.19 Centre of Torsion (CT) for a number of sections. CG = Centre of Gravity.

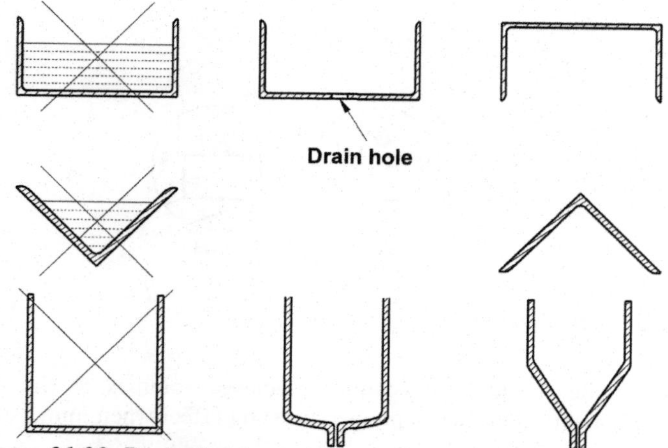

Figure 21.20 Designing to avoid corrosion.

Design to resist corrosion

Considerations can be made for corrosion at the design stage as follows:

- Design the structure to avoid corrosion.
- Design the structure so that it is easy to apply corrosion protection, and so that the corrosion protection can easily be maintained.
- Design the structure with an allowance for corrosion (rust allowance), i.e. so that some corrosion can occur without risk of failure or leakage.

It is important to avoid pockets and crannies in which dirt and water can collect, as shown in Figure 21.20. If it is not possible to avoid areas where water can gather, drain holes of at least 20 mm diameter should be provided in appropriate positions.

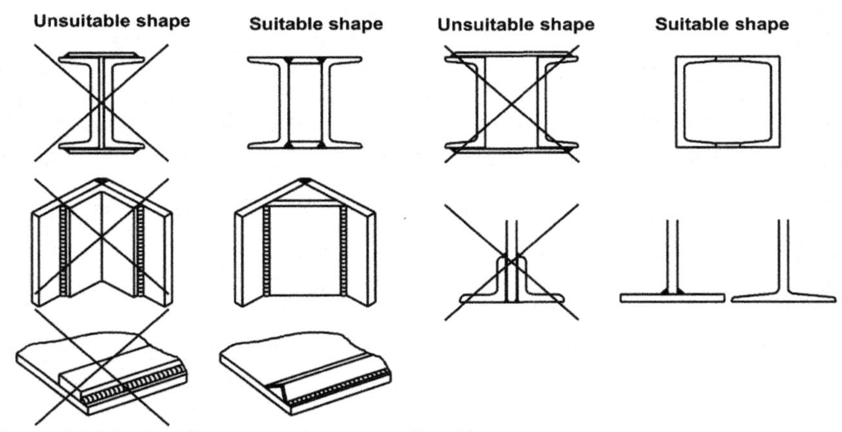

Figure 21.21 Avoiding corrosion around welds.

The structure should be designed so that narrow gaps are avoided, as shown in Figure 21.21. Use butt welds instead of overlap welds. If overlap welds cannot be avoided, they must be applied all the way round the material, taking particular care to avoid pores.

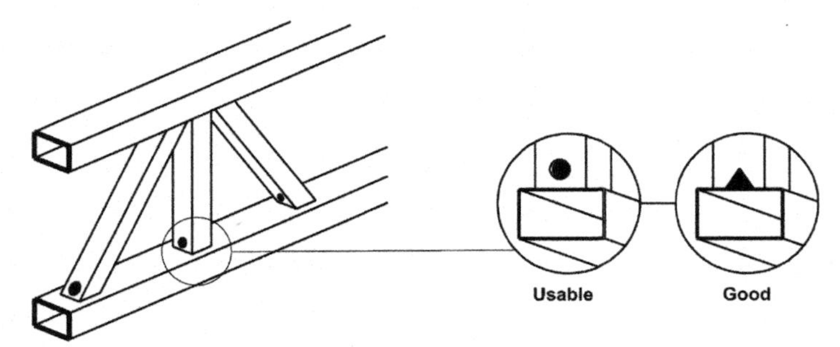

Figure 21.22 Designing for zinc-coating.

Rust protection can be applied by hot zinc-coating. Enclosed volumes, as shown in Figure 21.22, must have openings in order to prevent bursting effect when immersed in the solution of zinc.

Design for production

Product costs can be kept down, unnecessary work on the shop floor can be reduced and overall quality can be improved if, right at the initial sketch stage, the designer carefully thinks through the production aspects of the design, and understands how the particular production process operates. Some general points are as follows:

- Use standard rolled or extruded profiles, or steel castings, as far as possible, which will minimise the amount of welding required.
- Welding can be reduced, and the number of parts kept down, by bending sheet materials.

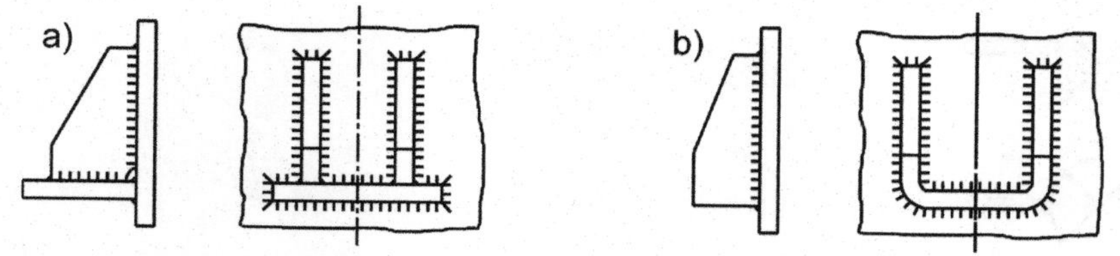

Figure 21.23 Welded and bent parts. Alternative b) is generally preferable.

- Use rational welding methods, such as spot welding, arc welding, friction welding etc.
- Consider the accessibility of parts to be welded (and accessibility for inspection and maintenance).
- Try to position joints so that the required welding position is comfortable. Horizontal welding is preferable to overhead welding.
- Choose a suitable groove, having the minimum filler metal needed to meet the necessary requirements in respect of quality, penetration requirements etc.
- Optimise the throat thickness of fillet welds. Doubling the throat thickness requires four times as much filler metal. Welding distortions also increase with increasing throat thickness. In general, some degree of penetration can be expected for fillet welds, which means that the actual throat will be somewhat thicker than as shown on the drawing.
- Use intermittent welding wherever possible.
- Select materials familiar to the manufacturer. High-strength materials can be difficult to handle and work.
- Try to avoid using too many different material qualities, sheet thicknesses or types of profiles, in order to avoid any risk of mix-ups.
- Specify the geometry, quality classes, inspection requirements etc. of welds unambiguously and in an optimum manner, in order to keep down manufacturing costs.
- Special consideration may be required if production is automated. The weld, for example, must be positioned accurately, and it may be necessary to provide greater space for the robot to reach the weld. Butt welds can be difficult if backing is not provided. Permissible tolerance levels are reduced.
- Using symmetrical welds can reduce welding distortions.
- Welding is made easier if the parts are self-locating and self-fixing.

Detail design

A good design for a welded structure is one in which it is easy to see how loads and forces are transferred and flow throughout the structure. In general, an item is well-designed if it is easy to calculate its strength. The following are a number of tips and views on appropriate design solutions.

- Avoid welding thin materials to thick materials (see Figure 21.24). This is undesirable in strength terms because of the resulting stress concentrations for fatigue loads, and in manufacturing terms due to the fact that the rapid cooling caused by the thick material can cause cracks and poor fusion in the welds.

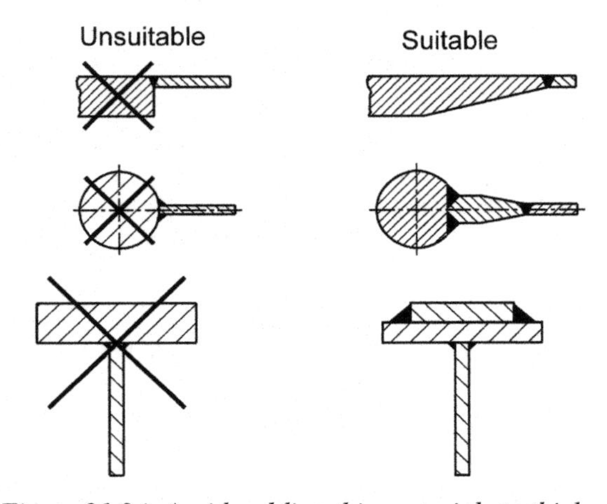

Figure 21.24 Avoid welding thin materials to thick materials.

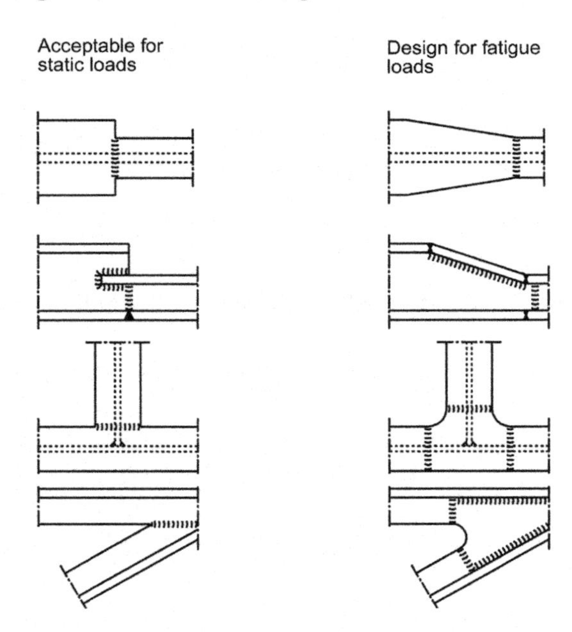

Figure 21.25 Design of changes in section.

- If the structure will be subject to fatigue loads, changes in sections must be designed so that there is as little stress concentration as possible, as shown in Figure 21.25.
- If possible, position welds in low-stress areas. Avoid positioning welds too closely to each other. Avoid also restraining welds (see Figure 21.26).
- Remember that it must be possible actually to produce the welds (Figure 21.27).
- Design joints as welded joints, and not as riveted joints with reinforcements (Figure 21.28).

Corners can be reinforced as shown in Figure 21.29. Figure 21.30 shows other examples of welded corners and corner reinforcements.

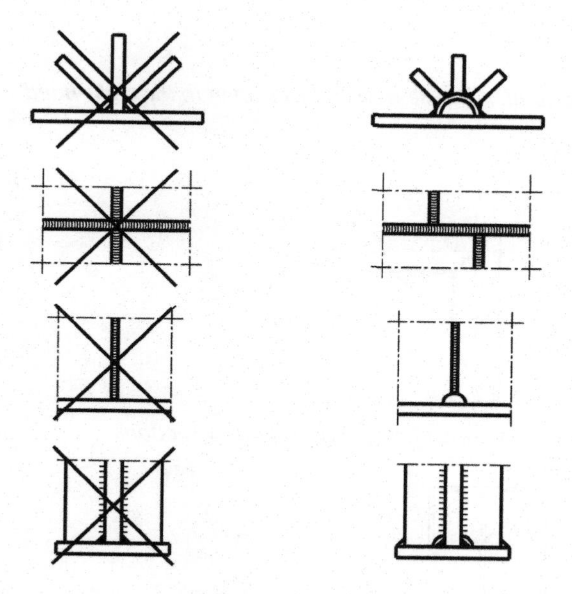

Figure 21.26 Avoid placing too many welds too close to each other.

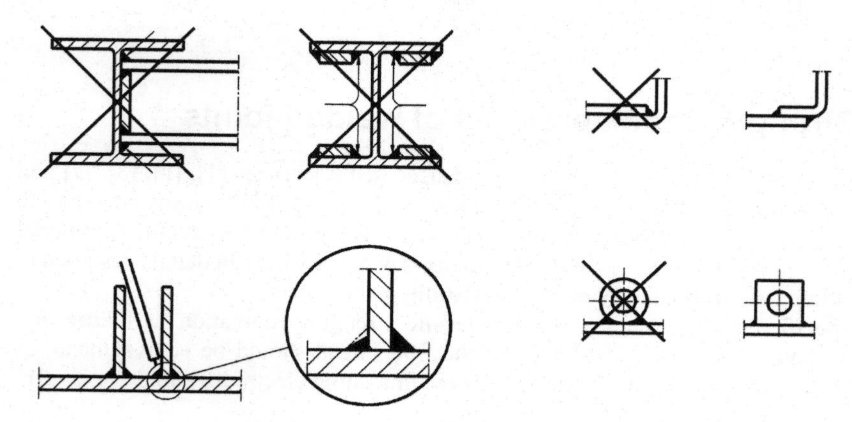

Figure 21.27 Remember to allow for access for welding.

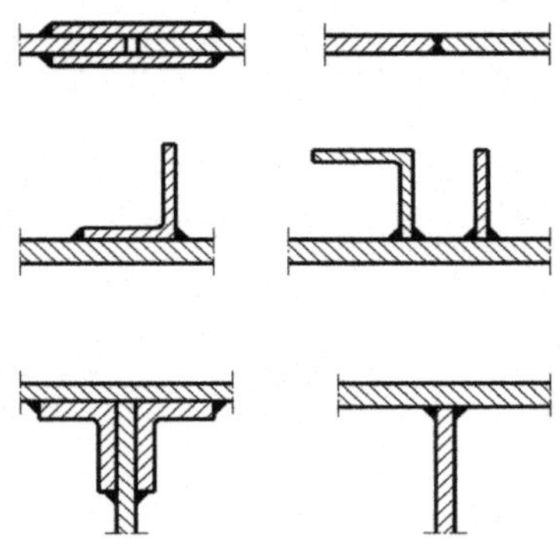

Figure 21.28 Riveted joint/welded joint. Choose a design as shown in the right-hand figures.

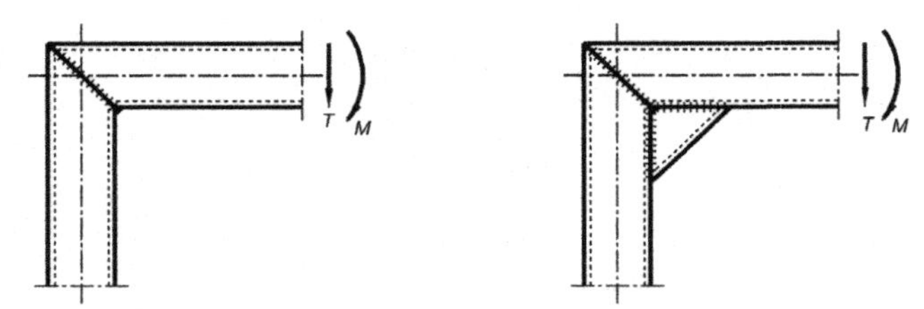

Figure 21.29 Reinforcing a corner consisting of rectangular hollow sections.

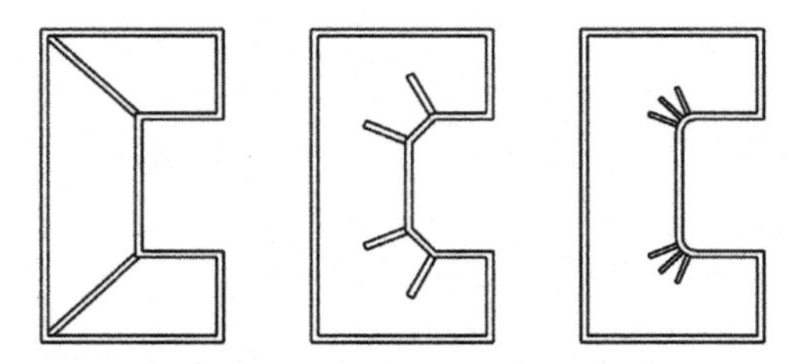

Figure 21.30 Corner reinforcements.

21.5 Strength considerations of welded joints

Section 21.5–21.7 is intended merely for guidance: see references [1], [2], [3], [7] and [8] for more detailed information on actual welded joints. If the component has to comply with any design code, then that code has to be followed in detail. Additional consideration to residual stresses in welded joints is necessary only in details designed to resist buckling, lateral instability or other instability phenomena.

The specified yield strength, ultimate tensile strength, elongation at failure and minimum Sharpy V-notch energy value of the filler metal should be equivalent to, or better than the parent material (= matching or overmatching electrodes should be used).

Butt welds
If the filler metal is overmatching, and if the weld is fully penetrating, then the welded joint will be stronger than the parent metal.

Fillet welds
The section for analysis is through the throat thickness of the weld, as shown in Figure 21.31.

In general, stresses can be assumed to be uniformly distributed along the analysed section. Some of the penetration might be included in the throat thickness [1].

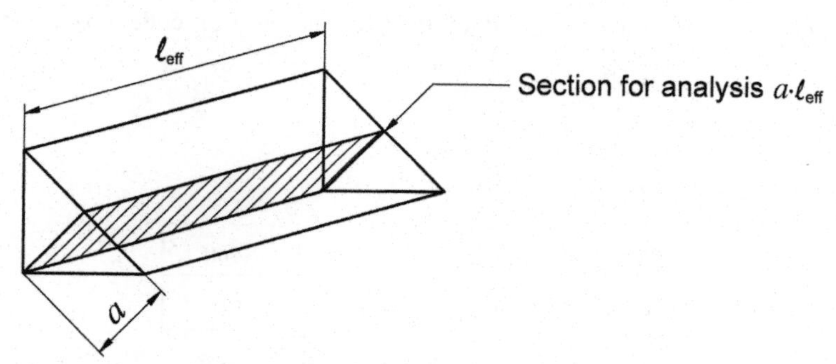

Figure 21.31 Section for analysis of fillet welds.

21.6 Analysis of statically loaded welded joints

A fillet weld can be subjected to the following stresses (see Figure 21.32):

$\sigma_{\parallel}$ = normal stress parallel to the weld (can be neglected in statically loaded structures).
$\sigma_{\perp}$ = normal stress perpendicular to the weld.
$\tau_{\parallel}$ = shear stress parallel to the weld.
$\tau_{\perp}$ = shear stress perpendicular to the weld.

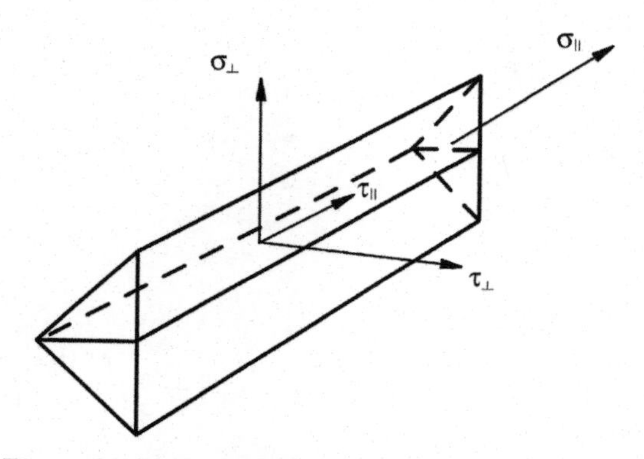

Figure 21.32 Stresses in a fillet weld.

Ultimate strength capacity

The design resistance of a fillet weld will be sufficient if the following are both satisfied:

$$\sqrt{\sigma_{\perp}^2 + 3 \cdot \left(\tau_{\perp}^2 + \tau_{\parallel}^2\right)} \leq \frac{f_u}{\beta_w \cdot \gamma_{M2}} \quad \text{(equation 1)}$$

$$\sigma_{\perp} \leq \frac{0.9 \cdot f_u}{\beta_w \cdot \gamma_{M2}} \quad \text{(equation 2)}$$

where f_u = the ultimate strength of the weaker part.
β_w = correlation factor according to Table 21.1.
γ_{M2} = partial factor for welded joints. According to Eurocode 3 it can be chosen to
$\gamma_{M2} = 1.25$

TABLE 21.1 Correlation factor β_w.

Material	S235	S275	S355	S420	S460
β_w	0.8	0.85	0.9	1.0	1.0

Example of calculations for a statically loaded joint

Calculate the capacity perpendicular to the weld in Figure 21.33.

Length of the weld, L = 200 mm.
Material: EN 10025: S275JR (f_u = 410 N/mm^2)
Electrode class ISO E42 4 B (f_{eu} = 500–640 N/mm^2)

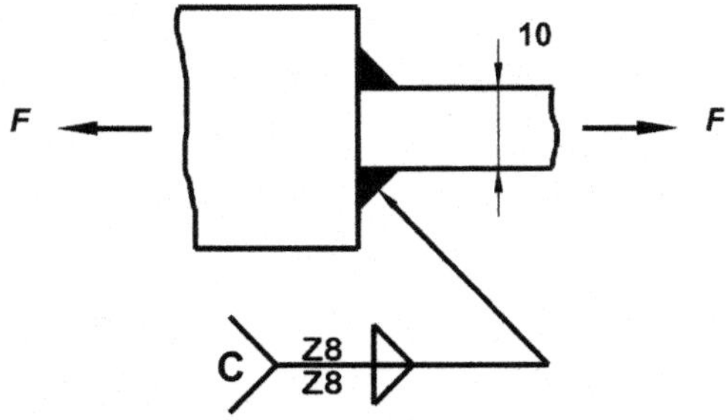

Figure 21.33 Double fillet weld.

Solution:

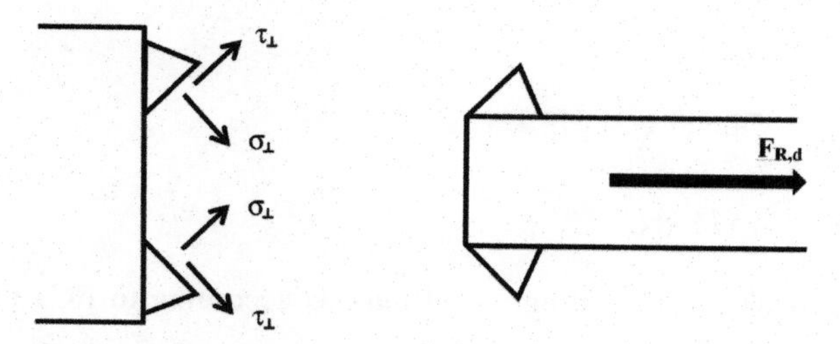

Figure 21.34 Stresses in fillet welds when the load is perpendicular to the weld.

One weld is carrying half the load:

$$\frac{F_{R.d}}{2} = \frac{\sigma_\perp \cdot a \cdot \ell}{\sqrt{2}} + \frac{\tau_\perp \cdot a \cdot \ell}{\sqrt{2}}$$

If $\sigma_\perp = \tau_\perp$ then

$$\frac{F_{R.d}}{2} = \frac{2 \cdot \sigma_\perp \cdot a \cdot \ell}{\sqrt{2}}$$ and finally

$$\sigma_\perp = \tau_\perp = \frac{F_{R.d}}{\sqrt{2} \cdot 2 \cdot a \cdot \ell}$$

To fulfill equation 1:

$$\sqrt{\sigma_\perp^2 + 3 \cdot \left(\tau_\perp^2 + \tau_\parallel^2\right)} \leq \frac{f_u}{\beta_w \cdot \gamma_{M2}}$$

In our case:

$$\sqrt{4 \cdot \sigma_\perp^2} \leq \frac{f_u}{\beta_w \cdot \gamma_{M2}} \quad \text{or} \quad \sigma_\perp \leq \frac{f_u}{2 \cdot \beta_w \cdot \gamma_{M2}}$$

Thus:

$$\frac{F_{R,d}}{\sqrt{2} \cdot 2 \cdot a \cdot \ell} \leq \frac{f_u}{2 \cdot \beta_W \cdot \gamma_{M2}}$$

Finally

$$F_{R,d} \leq \frac{f_u \cdot \sqrt{2} \cdot a \cdot \ell}{\beta_W \cdot \gamma_{M2}}$$

With $a = 8/(\sqrt{2}) = 5.6$ mm, $\ell = 200 - 2 \cdot 5.6 = 188$ mm, $f_u = 410$ N/mm^2, $\beta_w = 0.85$ and $\gamma_{M2} = 1.25$:

$$F_{R,d} \leq \frac{410 \cdot \sqrt{2} \cdot 5.6 \cdot 188}{0.85 \cdot 1.25} = 574\ 500\ \text{N} = 574\ \text{kN}$$

Answer: The load capacity is 574 kN.

Comments: The load capacity of the 10 mm steel plate is according to EC3 the lesser of the two values:

$$F_{R,dpl} \leq \frac{B \cdot t \cdot f_y}{\gamma_{M0}} = \frac{200 \cdot 10 \cdot 275}{1.0} = 550\ 000\ \text{N} = 550\ \text{kN} \quad \text{or}$$

$$F_{R,dpl} \leq \frac{0.9 \cdot B \cdot t \cdot f_u}{\gamma_{M2}} = \frac{0.9 \cdot 200 \cdot 10 \cdot 410}{1.25} = 590\ 000\ \text{N} = 590\ \text{kN}$$

In other words, the 10 mm steel plate will fail approximately at the same load as does the weld.

A simple rule of thumb for double fillet welds is that the leg length should be $0.8 \cdot t$ to $1.0 \cdot t$, where t = the thickness of the plate.

21.7 Welded structures subjected to fatigue loads

Some general notes on fatigue of welded joints

With steadily increasing demands for low weight and reduced manufacturing costs, it has become essential to solve the problem of welded structures subjected to fatigue loads. The notes on the following pages are based almost entirely on full-scale tests of structures, and the resulting experience of their performance. In order to understand the phenomenon of fatigue, it is essential to be aware of the following.

- *The geometry of the weld determines its fatigue strength.*

Examination of a welded joint under a microscope shows that there are always discontinuities, particularly in the transition between the parent metal and the weld. Defects can be such as undercutting, overlaps, overrunning welds, micro-cracks, excessive convexity etc. Together with these 'micro-geometric' potential stress concentrations, the 'macro-geometric' stress concentrations for example in the form of such features as sudden changes in the cross sections, determine the fatigue strength of the welded joint. With a little experience and some imagination, it is soon quite easy to see where fatigue cracks might be likely to occur in welded joints. Some examples are shown in Figure 21.35.

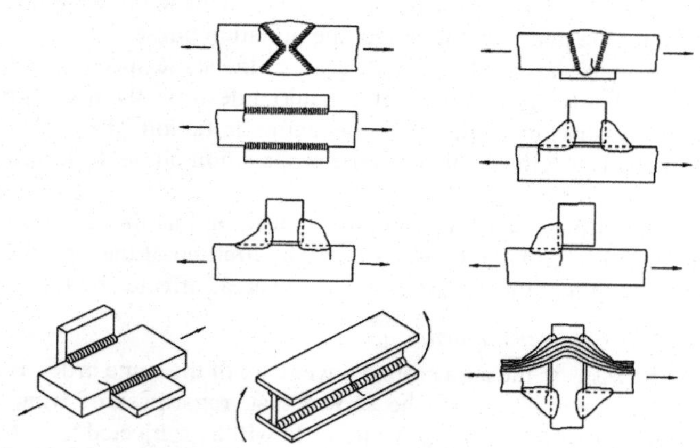

Figure 21.35 Critical points in a welded joint. Possible areas where cracks can be expected are marked.

Smoother transitions between welds and the parent metal can be obtained by grinding the weld or by TIG dressing it. In general, grinding is preferable for butt welds, while TIG dressing is often preferable for fillet welds.

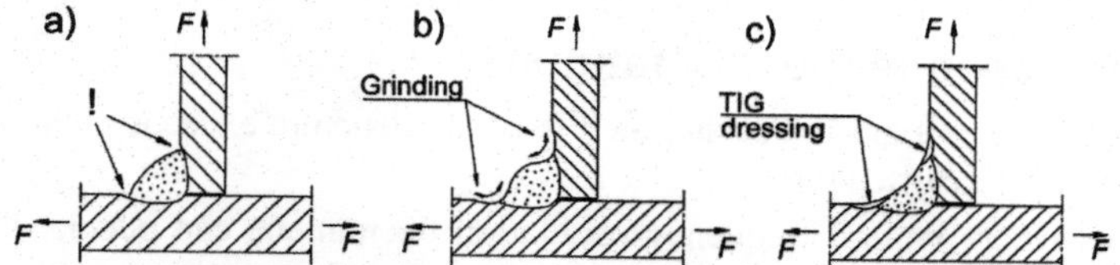

Figure 21.36 Grinding or TIG dressing reduces potential stress concentrations.
a) Before treatment. b) After grinding. The direction of grinding has been marked.
c) After TIG dressing.

It is particularly important to grind away any potential stress concentration effects in the transition area between the weld and the parent metal. To be on the safe side, this involves also grinding slightly into the parent metal. The reduction in stress concentration effects normally by far compensates the effect of the reduced thickness of the parent metal.

- *The static strength of the parent metal (and of the filler metal) is of less importance in determining the fatigue strength of a welded joint.*

This phenomenon can be explained by the theory of fracture mechanics. We normally distinguish between three different phases in the growth of a fatigue crack through the material:

Phase 1: *Crack initiation* = the number of load cycles required before the discontinuities form a real crack (of the order of hundredths of a millimetre in size).

Phase 2: *Crack propagation* = after formation, the crack grows by a given particular length for each new load cycle.

Phase 3: *Final failure* = when the crack has grown so much that the remaining material is unable to carry the applied static load, and so a final failure occurs.

The differences in the fatigue strengths of different materials depend on the length of the initiation phase: the better the material, the longer the initiation phase.

However, once a fatigue crack has started, it will grow at the same speed in practically all grades of steel. Welding (or gas cutting) of a material destroys the microstructure and introduces so many discontinuities that the entire initiation phase can be regarded as having been passed. All that remains is the propagation phase, followed by the final failure.

As the propagation phase is the same for all types of steel, and as the final failure phase has only marginal importance, it can be shown that the lifetime of the structure is independent of the fatigue strength of the material – that is, in areas affected by the weld.

- *The absolute stress level is of lesser importance.*

In general, the residual stresses in the area around a weld are of the same order as the yield tensile strength of the material. It can be shown that, regardless of what the maximum/minimum external stress ($\sigma_{max}/\sigma_{min}$) is, the weld will be subjected to a stress that varies from the yield tensile strength value down to the yield limit minus (σ_{max} - σ_{min}), see reference [3].

A decisive factor in determining the fatigue strength of a weld is thus the stress range $\sigma_r = \sigma_{max} - \sigma_{min}$. Note, however, that a welded joint that has been stress-relieved, or material that is not affected by the weld, is dependent on the absolute level of stress, as there are then lower residual stresses due to welding to add to the external stresses.

Some general advice in respect of fatigue

- If the number of load cycles expected during the life of the structure exceeds 1000, a fatigue analysis should be carried out.
- Be wary of designs where natural frequency oscillation phenomena can occur. In such cases, the number of load cycles will rapidly become very large.
- If the material is not affected by the weld, then the fatigue strength will be approximately proportional to the ultimate tensile strength of the material.
- If the metal has been thermally cut, it must be regarded in the same way as a welded element as far as fatigue assessment is concerned.
- If the structure is exposed to a corrosive environment, fatigue strength will be drastically reduced: in the worst case, by up to 40 %.
- Although high-strength material may be used, a higher fatigue strength cannot be assumed if:
  - the material is affected by the weld
  - the material has been thermally cut
  - the environment is strongly corrosive.

In all these cases, fatigue strength higher than what would be the case for normal mild steel should not be expected. Nevertheless, high-strength steels can still be used if welds and thermally cut surfaces can be positioned in low-stressed areas (e.g. along the neutral axis of a beam subject to a bending moment), or if other special measures can be taken.

No	Joint configuration	FAT	Remarks
01	Parent material, grounded surface	215	
02	Parent material, rolled surface	175 156 140	Good quality Moderate quality Exposed surface of slow rusting steel
03	Parent material, sand blasted surface		See Nos 02 and 04 – 07
04	Parent material, hot dip zinc coated surface	156	
05	Parent material, sawn surface	156 175	Workmanship Class GB Workmanship Class GA Chamfered edges
06	Parent material, sheared surface		See No 05
07	Thermally cut surface	90 112	Surface roughness < 0.3 mm Corners machined to 2 mm chamfer. Surface roughness < 0.2 mm

No	Joint configuration	$FAT_{//}$	$FAT_{\perp}$	Remarks
10	Butt weld in double V joint 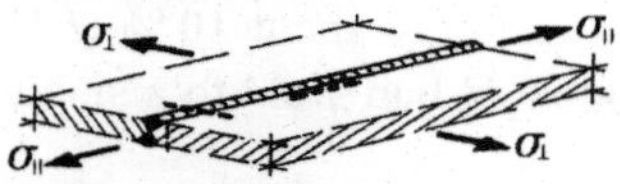	100	90	
11	Butt weld in single V joint 	100	90	With root with a sealing run, alternatively welded against a backing strip which is removed
12	Butt weld in single V joint 	90	71	No sealing run on root, but the quality requirement for the specified weld quality level shall apply for the root side also
13	Butt weld in single V joint with backing strip left in position 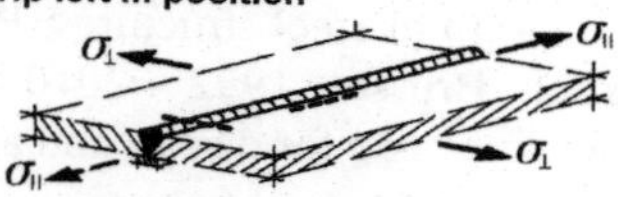	90	56	
14	Butt weld with incomplete penetration 	90	-	

Figure 21.37 Examples of joint classes. Quality as per welding class B according to SSAB Steel Sheet Handbook.

Hot zinc-coating reduces the fatigue strength of high-strength steels. However, this effect is negligible if the ultimate tensile strength of the basic material is less than 500 N/mm^2. For a UTS of 800 N/mm^2, there can be a reduction of up to 30 % in relation to the strength of an untreated rolled surface. Nevertheless, in a strongly corrosive environment, hot zinc-coating is still preferable to the use of untreated material.

Joint classes

Over the years, considerable knowledge and experience of fatigue testing of parts and materials has been built up. One way of bringing together such material is to create a 'reference library' of tested parts. Most standards and regulations put welded joints into various categories or classes (FAT). See, for example, references [1], [2], [3] and [7]. The joint class specify the characteristic fatigue strength of the weld in N/mm^2 for 2×10^6 load cycles.

A reference library of joint classes can be very valuable, even without performing any fatigue analyses or any other calculations at all. Figure 21.8 is taken partly from reference [3].

Example A: Study joint 10 (Double-V butt weld). $FAT_{\parallel} = 100$ and $FAT_{\perp} = 90$ N/mm^2. This weld is therefore about 10 % stronger in respect of loads parallel to its longitudinal direction compared to loads perpendicular to the weld.

Example B: Then compare joint 13 (butt weld with backing strip left in position): $FAT_{\parallel} = 90$ and $FAT_{\perp} = 56$ N/mm^2.

If the applied stress is parallel to the weld, we lose only about 10 % ($FAT_{\parallel}$ being reduced from 100 to 90) when changing from a double-V butt weld to a single-V butt weld, welded only from one side.

If, instead, the stress is perpendicular to the weld, we lose almost 40 % of its strength ($FAT_{\perp}$ falling from 90 to 56). This result feels right, as the backing strip must have a greater stress concentration effect perpendicular to the weld compared to if the stresses were acting parallel to the weld.

21.8 References and further reading

The following material can be particularly recommended for those wishing to learn more about dimensioning and design of welded products.

[1] ENV1993-1-1 to ENV1993-1-12 Eurocode 3: Design of steel structures. Part 1-1 to 1-12: European Committee for Standardisation. Brussels 1992 – 2010. (European design code for steel structures. Comprehensive code for design and analysis of buildings and constructions in steel).

[2] Recommendations for fatigue design of welded joints and components. A Hobbacher, Welding Research Council, Bulletin 520. New York 2008. ISBN 1 58145 527 5. (Tips, advices and instructions for constructive design and dimensioning of welded components not covered by specific sector standards and submitted to fatigue loads).

[3] The Steel Sheet Handbook, SSAB Tunnplåt AB. Edition 3, 1996. (A general handbook for design and dimensioning of thin sheet structures in high-strength materials).

[4] ISO 2553. Welded, brazed and soldered joints. Symbolic representation on drawings. Edition 1, 1994-09-16.

[5] ISO 5817. Arc-welded joints in steel. Guidance on quality levels for imperfections. Edition 1, 1993-02-26.

[6] Analysis of Welded Structures, Koichi Masubuchi, Pergamon Press Ltd, London, 1980. ISBN 0-08-022714-7.
(A guide to estimating the magnitude of distortions and residual stresses).

[7] BS 7608:1993 Code of practice for Fatigue design and assessment of steel structures. British Standards Institution, London 1993. ISBN 0 580 21281 5.

[8] EN 13445-3 Unified pressure vessels – Part 3: Design. ICS 23.020.30. European Committee for Standardisation, Brussels 2002. (Comprehensive handbook for design and analysis of pressure vessels).

Author: Claes Olsson

22 Quality assurance and quality management

22.1 Introduction

Quality requirements in respect of welded structures are set out in directives, regulations, standards or customer specifications. The manufacturer performing the work needs to analyse these requirements at the tendering stage, to decide whether or not they can be fulfilled. This is assisted by a systematic method of working. Companies with ISO 9001 certification have documented procedures for this. ISO 9001 is a system standard concerned with quality systems. They define welding as a special process that must be properly controlled in order to ensure that the necessary quality requirements are fulfilled.

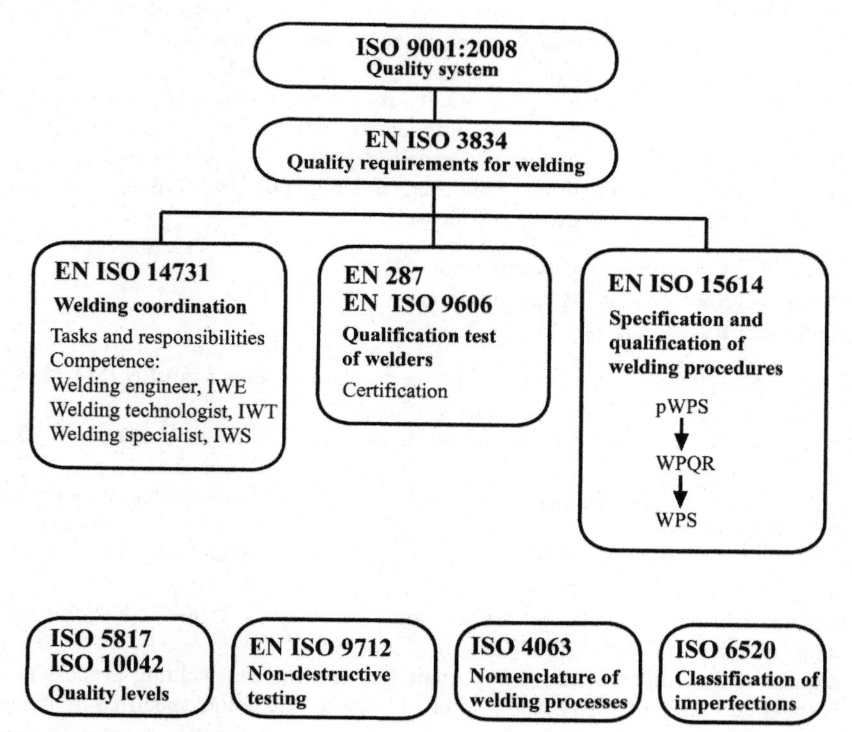

Figure 22.1 Standards that regulate quality requirements for welded structures.

Figure 22.1shows standards that are of interest in this context. The following are further described below:

- EN ISO 3834, Quality requirements for welding
- EN ISO 14731, Welding coordination – tasks and responsibilities

- EN 287, Qualification test of welders
- EN ISO 15614, Specification and qualification of welding procedures

22.2 Quality requirements for welding (EN ISO 3834)

EN ISO 3834 is a process standard for welding, and describes how quality assurance of welding work can be ensured. It consists of sex different parts. EN ISO 3834-1, which is a guideline part, sets out the following application areas:

a) as guidance for specification and establishment of that part of ISO 9001 concerned with the management of special processes.
b) as guidance for determination of welding quality requirements in those cases where the quality system standards are not applicable.
c) in connection with auditing/assessment of welding quality in accordance with a) and b) above.

The standard has three different quality requirement levels, relating to *comprehensive requirements, standard requirements* and *elementary requirements*, so that the supplier/customer/requirements specifier can choose the requirement level that is suitable for the welding work to be performed.

EN ISO 3834-2 is used for all three quality requirement levels when ISO 9001 requirements apply. This is because the requirements in EN ISO 3834-2 can be set at a suitable level for the particular structure concerned, depending on the effect of welding on the product safety and function. However, if ISO 9001 requirements are not involved, then EN ISO 3834 is applied, as follows:

- EN ISO 3834-2 Comprehensive quality requirements
- EN ISO 3834-3 Standard quality requirements
- EN ISO 3834-4 Elementary quality requirements

The next stage in the process is to select the particular elements from EN ISO 3834-2, -3 or -4 that are applicable to the particular working area.

SS-EN ISO 3834-5 contains references that need to agree with the demands in SS-EN ISO 3834-2; -3; or -4.

EN ISO 3834-6 gives guidelines for the implementation of requirements given in the other parts of EN ISO 3834.

EN ISO 3834-4 specifies a minimum acceptable quality requirement level, from which no elements may be excepted. If there are requirements in respect of general quality management systems, EN ISO 3834-2 must be chosen. Table 22.1 shows the requirements included in EN ISO 3834-2.

The requirement elements that are important for ensuring the welding process are number 7, together with parts of 10, 11, 12 and 13. Among the points specified by them are:

- that welders must have been tested and qualified in accordance with EN 287 (Section 7.2), EN 1418, EN-ISO 9606-2, -3, -4, -5.
- that NDT personnel must be qualified and approved (Section 7.2).
- that welding must be carried out in accordance with a welding procedure specification (WPS) (Section 10.2)

- that consumables must be stored and handled in such a way as to prevent absorption of moisture (Section 11.3)
- that the parent materials must be handled so that they cannot be confused with other materials (Section 12)
- that the need for heat treatment must be considered (Section 13)

TABLE 22.1 Requirement elements in EN ISO 3834-2.

Requirement element	Heading number in the standard
Review of requirements and technical review	5
Sub-contracting	6
Welding personnel	7
Inspection and testing personnel	8
Equipment	9
Welding and related activities	10
Welding consumables	11
Storage of parent materials	12
Post-weld heat treatment	13
Inspection and testing	14
Non-conformance and corrective actions	15
Calibration and validation of measuring, inspection and testing equipment	16
Identification and traceability	17
Quality records	18

The differences between EN ISO 3834-2 and EN ISO 3834-3 are slight, and relate primarily to requirements in respect of equipment maintenance, calibration, qualification of WPS and batch inspections of electrodes. EN ISO 3834-1 includes an appendix that provides an overall description of the differences between the three levels of the standard.

EN ISO 3834 specifies requirements in respect of a welding coordinator, and refers to EN ISO 14731, Welding Coordination – Tasks and Responsibilities (see Section 22.3).

The quality requirements specified in respect of welded products can be verified as follows:

- Specification and qualification of welding procedures, which verifies the mechanical properties of the welded joint (EN 288, see Section 22.4).
- Qualification testing of welders, which verifies the competence of the welder/welding operator (EN 287, see Section 22.5, EN 1418, EN-ISO 9606-2, -3, -4, -5).
- Non-destructive testing, which verifies that the welded joint does not contain impermissible imperfections (ISO 5817, see Section 22.6).

22.3 Welding coordination (EN ISO 14731)

Welding is a process that requires management and coordination in order to ensure that the specified quality requirements can be fulfilled. EN ISO 14731 describes the duties and responsibilities associated with such coordination and management of welding, briefly summarised below.

The extent of the coordination required depends on the manufacturer's own requirements, requirements in applicable standards and requirements in the contract. The duties in connection with this coordination and management can be shared by a number of persons. However, it must be defined, e.g. by documents describing the duties of the persons concerned.

TABLE 22.2 Requirement elements in ISO 9001 in comparison with EN ISO 3834 -2, -3 and -4. The figures refer to the heading numbers in the standards.

ISO 9001		EN ISO 3834-2	EN ISO 3834-3	EN ISO 3834-4
4.1.2	Organisation	6.1	6.1	
4.1.2.2	Personnel and equipment for verification	7.1/2	7.1/2	-
4.3	Contract review	4.2	4.2	(4)
4.4.5	Design review	4.3	4.3	(4)
4.6	Purchasing	5	5	(5)
4.8	Product identification and traceability	16	(15)	-
4.9	Process control – planning	9.1	(9.1)	-
	-"-- instructions	9.2/4	(9.3)	(8)
	-"-- welding procedure qualification	9.3	9.2	-
	-"-- workshop capacity	8.1/2	(8.1/2)	-
	-"-- equipment	8.3/4	(8.3)	-
	-"-- maintenance	8.5	(8.3)	-
	-"-- heat treatment	12	(12)	-
4.10.2	Process control and testing	13.2/3	13.2/3	
4.10.3	Final inspection and testing	13.4	13.4	
4.10.4	Inspection and test reports	9.5	-	-
4.11	Calibration	15	-	-
4.12	Inspection and test status	13.5	13.5	-
4.13	Non conformance products	14	(14)	(11)
4.14	Corrective actions	14	(14)	(11)
4.15.2	Handling	10.3	10.2	-
4.15.3	Storage	11	11	-
4.16	Quality documents	17	16	(12)
4.18	Training	6.2/3	6.2/3	6

( ) = Less extensive requirements
– = No requirements

Examples of such duties include:

- specification
- control/coordinate
- inspect/witness

Somebody must be appointed as the manufacturer's authorised *welding coordinator*, authorised to issue/approve the necessary welding documents on behalf of the manufacturer.

Of the activities listed in EN ISO 3834, the following can be linked to quality-related duties in accordance with EN ISO 14731:

- General requirements
- Technical requirements
- Purchasing of parent materials
- Selection of consumables
- Selection of subcontractors
- Production planning
- Qualification of welding procedures
- Selection of equipment
- Qualification of welders
- Personnel for inspection and testing
- Performing the welding
- Inspection and testing
- Documentation

A suitable way of meeting the above requirements in connection with the manufacture of welded products is to use checklists for the preparation of tenders and for production planning.

Qualifications

Welding coordinators must possess the necessary qualifications for their duties, in the form of general and specialized technical knowledge, which is attained by a combination of theoretical knowledge, training and/or experience from the welding industry. Under EN ISO 14731, technical knowledge can be divided into three levels: comprehensive, standard and basic.

Examples of training and qualifications that are regarded as fulfilling the requirements in respect of technical knowledge are the following, as issued by the International Institute of Welding, IIW:

- International Welding Engineer, IWE
- International Welding Technologist, IWT
- International Welding Specialist, IWS

22.4 Specification and qualification of welding procedures

Specifications for and qualification of welding procedures for welding metallic materials. It consists of the standards in Table 22.3.

General rules (EN ISO 15607)

Application

This section of the standard defines general rules for description and qualification of welding procedures. It assumes that welding will be carried out using conventional welding methods, controlled by a welder or welding operator working in accordance with a welding procedure specification or welding data sheet. The standard applies when qualification of welding procedures is called for, e.g. in contracts, product standards, regulations or directives.

TABLE 22.3 Standards for qualification of welding procedures.

SS-EN ISO 15607	General rules
SS-EN ISO 15609-1	Welding procedure specifications, arc welding
SS-EN ISO 15609-2	Welding procedure specifications, gas welding
SS-EN ISO 15609-5	Welding procedure specifications, resistance welding
SS-EN ISO 15610	Qualification based on tested welding consumables
SS-EN ISO 15611	Qualification based on previous welding experience
SS-EN ISO 15612	Qualification by adoption of a standard welding procedure
SS-EN ISO 15613	Qualification based on pre-production welding test
SS-EN ISO 15614-1	Welding procedure test - Arc and gas welding of steels and arc welding of nickel and nickel alloys
SS-EN ISO 15614-2	Welding procedure test - Arc welding of aluminium and its alloys
SS-EN ISO 15614-5	Welding procedure test - Arc welding of titanium, zirconium and their alloys
SS-EN ISO 15614-6	Welding procedure test - Arc and gas welding of copper and its alloys
SS-EN ISO 15614-12	Welding procedure test - Spot, seam and projection welding

Specification of welding procedures

All welding operations must be sufficiently planned before production starts. This includes producing welding procedure specifications for all welded joints, in accordance with the requirements of EN ISO 15609, and providing as much detail as required by the qualification method. All important variables that could affect the properties of the welded joint must be included. Any permissible variations must be specified.

Until the welding procedure specification has been qualified, it is classified as *preliminary*, pWPS.

Welding procedure specification (EN ISO 15609)

EN ISO 15609 specifies the technical contents of the welding procedure specification (WPS) for arc-, gas- and resistance welding methods.

A WPS must specify, in detail, how welding is to be performed. It must contain all important information relating to the welding work, with indication of whether such factors can affect the metallurgy, mechanical properties or geometry of the welded joint.

The nomenclature of welding and allied processes is specified and numbered in ISO 4063. Numbers designating the most common welding methods are shown in the table below. Model forms for WPS are given in the standard in the form of an appendix.

Welding procedure tests for arc and gas welding of steel and arc welding of nickel and nickel alloys (EN ISO 15614-1)

EN ISO 15614-1 sets out the conditions for welding procedure tests on standardised test pieces for the arc welding of steel, and includes the welding methods in Table 22.4.

TABLE 22.4 Numerical reference numbers of common fusion welding methods as given in ISO 4063.

Welding method	ISO 4063 designation
Metal arc welding with covered electrode	111
Flux cored metal-arc welding without gas shield	114
Submerged arc welding	12
MIG-welding	131
MAG-welding	135
MAG-welding with flux cored wire	136
TIG-welding	141
Plasma arc welding	15
Oxy-acetylene welding	311

Other fusion welding methods can be included, subject to agreement between the parties.

Test pieces

The standard specifies the shape and minimum dimensions of standardised *test pieces* to be used in connection with the welding procedure. The test pieces must be sufficiently large to ensure that there is sufficient material to conduct away the heat. When impact testing of the heat-affected zone is required, the test pieces must be marked with the rolling direction.

All welding of test pieces must be carried out in accordance with the preliminary WPS, and under the same conditions as can be expected in production. Welding positions, angles of slope and rotation must be as specified in EN ISO 6947. Tack welding must be included in the test welds if it is to be used in production. Welding and testing must be supervised by an examiner or examining body.

Examination and testing

Testing consists of both non-destructive and destructive testing, as appropriate, and as follows:

- visual inspection
- radiographic or ultrasonic testing
- crack detection
- transverse tensile test
- transverse bend test
- impact testing
- hardness test
- macro and micro examination

The standard specifies how the test pieces shall be positioned.

Retesting

If the welding procedure test pieces do not meet all the test requirements, the results cannot be qualified. It is permissible to perform a further procedure test.

If any single test piece fails to meet the requirements due to geometrical defects, two new test pieces may be selected for retesting. If either of them fails, then the entire WPS also fails.

Range of qualification

A WPS that has been qualified by a manufacturer is valid for welding in workshops and at sites under the same technical management.

Welding procedure tests form the basis for qualification of a WPS, of which the important variables lie within the qualification range of the procedure test. Essential variables are:

- parent material
- material thicknesses
- welding method
- welding position
- type of joint
- consumables
- type of welding current
- heat input
- preheat temperature
- intermediate pass temperature
- post heat treatment

Documents of qualified welding procedure tests (WPQR)

Records from welding and testing shall include all the information needed for qualification. Welding Procedure Qualification Records (WPQR) must be signed by the examiner. Model forms of WPQR are included in the standard.

Older welding procedure tests

Older welding procedure tests, carried out in accordance with national standards or specifications, can be qualified provided that the technical requirements in EN ISO 15614 are fulfilled and that the test conditions correspond to the production conditions that will be encountered. Use of these older welding procedure tests shall be agreed between the contracting parties.

Welding procedure test for arc welding of aluminium and its alloys (EN ISO 15614-2)

EN ISO 15614-2 describes the conditions applicable to qualification of welding procedures to be used for arc welding of aluminium and its weldable alloys in accordance with ISO 2092 and 2107. These welding methods are MIG welding, TIG welding and plasma welding.

The standard, which follows the same principles as in EN ISO 15614-1, specifies how welding is to be performed and what tests that are to be carried out. Important variables for the procedure test are the same as for steel, but with lesser differences in the validity area. Note that welding position is an essential variable.

Tensile testing employs a correction factor linked to the type of alloy of the parent material and its as-delivered conditions. Bend testing is carried out using a larger former diameter for the high-strength alloys than for untreated aluminium.

Use of tested welding consumables (EN ISO 15610)

EN ISO 15607 allows welding procedures to be qualified on the basis of the use of tested consumables. This method is described in EN ISO 15610, which applies for repetitive welding operations and for workpiece materials of which the structures and properties in the heat-affected zone do not degrade during operation.

For steel, applicable welding methods are metal-arc welding, MIG/MAG welding and TIG welding, while MIG welding and TIG welding are applicable for aluminium. The standard applies to carbon manganese steels and chrome nickel steels, as well as for pure aluminium and non-heat-treatable aluminium alloys. Parent material thicknesses are 3–40 mm.

Qualification is given by an examiner or examining organisation, based on the workpiece material specification in accordance with an EN standard, and description of tested consumables in accordance with the relevant EN standards and a specific pWPS in accordance with EN ISO 15609. Qualification is valid as long as the tested consumables continue to be used, and is documented by means of the examiner's initials and dating on the pWPS concerned.

Qualification related to previous experience (EN ISO 15611)

Many workshops have considerable experience of the manufacture of welded structures involving third-party inspection, with good operating experience of the finished products. In such cases, the welding procedure can be qualified on the basis of reference to previous experience. EN ISO 15611 describes the conditions for this procedure, and covers metal arc, submerged arc, MIG/MAG, TIG and plasma welding.

It must be possible to document an EN ISO 15609 pWPS based on previous experience by authentic tests or investigations that show that the technical specification requirements for the product are fulfilled. Two methods of documentation are specified:

1. Documentation of testing (e.g. non-destructive or destructive testing, leak testing), together with a summary of welding production over a period of at least one year.

2. The performance records of welds in operation over a suitable period (five years can be suitable).

Range of qualification is in accordance with EN ISO 15614-1 and 15614-2, and continue to apply as long as manufacturing is carried out in the prescribed manner. The qualification is documented by the examiner initialling and dating the preliminary WPS, which is then kept by the manufacturer.

Qualification by a standard welding procedure (EN ISO 15612)

EN ISO 15612 describes the conditions for qualification and use of a standard welding procedure. These procedures are restricted to the material groups and workpiece thicknesses specified in EN ISO 15610.

A standard WPS must comply with the requirements in EN ISO 15609-1, and be qualified by an examiner or examining body who/that verifies that welding and testing are carried out in accordance with the requirements of the standard. Organisations that have prepared qualified standard welding procedures can then supply them as basis for other companies various welding data sheets. The use of standard welding procedures, as of tested consumables or previous experience, can be limited by standards applicable to particular products or by requirements in contracts.

The use of standard welding procedures requires the involvement of a welding coordinator in accordance with EN ISO 14731, coupled with a requirement that the manufacturer's quality management system must fulfil the requirements of the applicable part of EN ISO 3834.

Standard welding procedures are valid as long as the above requirements are fulfilled.

Qualification by a pre-production weld test (EN ISO 15613)

Welding procedures may be qualified by pre-production weld tests if the shape and sizes of test pieces in accordance with the standard do not represent the particular types of welded joints to be made. The conditions associated with this method of qualification are set out in EN ISO 15613.

The test pieces must comply with the applicable product standard, or be as agreed between the contracting parties. A pWPS must be prepared before welding the test pieces, which must be carried out under conditions representative of the planned production.

As far as possible, testing shall include the various requirements given in the standard. In general, the following tests must be carried out:

- visual inspection
- crack detection: see Section 22.6)
- macro examination
- hardness test (depending on the material requirements).

In general, the validity range of qualification is as set out in the applicable parts of EN ISO 15614, but restricted to the type of joint used for the testing. Qualification of the procedure remains valid as long as production conditions are the same as those used during testing.

As far as possible, WPQR must comply with EN ISO 15614-1 or 15614-2.

22.5 Qualification test of welders (EN 287-1)

In general, directives and regulations relating to welded products require the competence of the welder to be stated. This can most easily be done by testing the knowledge and capabilities of the welders. EN 287-1, sets out important requirements for qualification and certification of persons welding steel. Welding operators are tested in accordance with EN 1418: those welding aluminium, copper, nickel or titanium are tested in accordance with EN-ISO 9606-2, -3, -4 and -5.

Welder certificates issued by an examination body in one country must be acceptable to examining bodies in other countries.

Essential parameters for welder testing

Welding methods

EN 287-1 covers metal arc welding with covered electrodes, submerged arc welding, MIG/MAG welding, TIG welding, plasma and gas welding. Other fusion welding methods may be included, if specifically agreed.

Types of welds

Butt welds and fillet welds in sheet, plate and pipe (hollow sections are regarded as being pipes).

Material groups

Material groups are in accordance with CEN ISO/TR 15608 specified as

1	CMn steel, $R_{eH} < 460$ MPa
2	Thermo-mechanically controlled processed (TMCP) steel
3	Quenched and tempered steels, $R_{eH} > 360$ MPa
3	Fine-grained steel, normalized, quenched and tempered steel, $R_{eH} > 360$ MPa
4,5,6	CrMo/CrMoV creep resisting steels
7	Ferritic and martensitic stainless steel, Cr = 12–20%
8	Austenitic stainless steels
9	Nickel alloy steels, Ni ≤ 10%.
10	Austenitic ferritic stainless steels (duplex).
21	Pure aluminium
22	Non-heat-treatable aluminium alloys
23	Heat treatable aluminium alloys
31	Pure copper
32-38	Copper alloys
43	Pure nickel
51	Pure titanium
52-54	Titanium alloys
61	Pure zirconium
62	Zirconium alloys

Information on other important aspects of welder testing, such as consumables, metal thicknesses, pipe diameters and welding positions, is given in the standard.

Performing welder tests

Qualification test of welders in accordance with EN 287-1, 1418 and EN ISO 9606 means that the manufacturer must review its activities and decide what material qualities, metal thicknesses and pipe diameters the manufacturer needs to weld. Welding positions for the various welding methods must also be defined.

When deciding on the type of product (plate/sheet or pipe), the manufacturer must consider the requirements in the standard concerned with renewal of welder testing: with reasonable continuity, welders must have carried out welding work covered by their qualification.

In general, qualification of a welder's capabilities also includes qualification of all welds made by the welder and regarded as easier than the test weld. The validity of qualifications of welders is set out in tables in the standard.

Welder tests must be conducted in accordance with a welding procedure specification or welding data sheet. This may be preliminary (pWPS), based on the manufacturer's aggregated welding abilities, or qualified, depending on the particular product requirements.

Welder testing must be supervised by an examiner or an examining test body that is accepted by the contracting parties. The examiners may be employed by the manufacturer or purchaser, or by a third party, as determined by the contract requirements. In certain cases, an external independent examiner or examining body can be required.

Although not obligatory, the welder's theoretical knowledge may also be tested. This will require the welder to have knowledge of:

- The important parts of the welding equipment
- Identification of parent materials
- Methods for preheating and checking preheat temperatures
- Handling of consumables
- Welding procedure specifications, selection of welding parameters
- The causes of welding defects and measures to avoid defects
- Safety requirements
- Validity of the welder's certificate.

The test methods used for assessment are:

- Visual inspection
- Radiography
- Bend test
- Fracture test
- Macro examination.

The method to be used depends on the welding method, the shape of the product and the type of joint.

Acceptance criteria for test welds shall be determined in accordance with EN ISO 5817, Level B, for steel, and EN ISO 10042 for aluminium, with some exceptions. EN ISO 6520 describes imperfections. Reference should also be made to corresponding acceptance criteria for non-destructive testing.

Duration of validity

Welder certificates are issued under the full and sole responsibility of the examiner or examining body. They remain valid for two years, provided that the welder continues to work within the validity area and does not take a break of longer than six months. This must be certified by the employer/foreman every sixth month.

Certificates can be extended for further periods of two years, provided that the welder's work continues to fulfil specified quality requirements, which must be confirmed by documentation of tests of the welding work. The examiner or examining body who/that issued the certificate can extend the validity of the certificate if the above conditions are fulfilled.

22.6 Non-destructive testing

Quality levels must be clearly specified if the quality of welds is to be determined by non-destructive testing. This is done in accordance with ISO 5817 and ISO 10042, Fusion-welded Joints in Steel and Aluminium - Guidelines for Quality Levels for Discontinuities and Geometric Imperfections. These two standards have three different quality levels:

Quality level	**Symbol**
Moderate	D
Average	C
High	B

The quality level is determined by visual testing (VT) and non-destructive testing. The most widely used methods of non-destructive testing are:

Radiography	RT
Ultrasonic testing	UT
Magnetic particle testing	MT
Liquid penetrant testing	PT
Eddy current testing	ET

Radiography is used to reveal primarily internal volumetric discontinuities. The films are evaluated against the required quality level. Use of this method is limited by the thickness of the weld, which should not exceed 50 mm.

Ultrasonic testing is most suitable for detecting internal plane discontinuities. It requires a good test surface. The material should not be less than 8 mm thick for reliable evaluation.

Magnetic particle testing is a surface testing method, used to reveal discontinuities in or immediately below the surface of ferromagnetic materials.

Liquid penetrant testing is also a surface testing method, and is used to reveal discontinuities in the surface of non-porous materials.

Eddy current testing, or inductive testing, is used to reveal discontinuities on or immediately below the surface of electrically conducting materials.

A common feature of all these test methods is that the test personnel must hold certificates in accordance with EN ISO 9712. Testing must also be carried out in accordance with qualified test procedures.

Author: Curt Johansson

22.7 References and further reading

S Hughes, *A quick guide to welding and weld inspection*, Woodhead Publishing Limited, 2009

B. Raj, *Practical non-destructive testing* (Second edition), Woodhead Publishing Limited, 2001

23 Welding costs

23.1 Introduction

The choice of a particular welding process is usually made on cost as well as technical terms. Technical limitations can be such as the type of material, its thickness, the type of joint required and/or welding position, i.e. factors that are directly linked to the capabilities of the particular welding method. In addition, there are limitations that may be imposed by special quality requirements, production resources in the factory, the work environment and so on. Despite potential benefits, a company may not have been able to invest in new welding equipment or a new welding method, with all that is involved in terms of equipment cost, training of personnel and bringing the new facilities into production.

As a result, the general determining factor in deciding on a particular welding process is usually cost – that is, to choose a process that produces the required quality at the lowest possible cost. This is assisted by a method of making welding cost calculations by means of a refined internal cost analysis procedure that calculates the costs that are specific to the welding element of the work. Traditionally, these costs are regarded as consisting of those for labour, consumables (which include shielding gas and backing materials), equipment and energy. There are, of course, many other costs, such as those for joint preparation, inspection, painting etc., all of which must be considered. The following material considers only the costs connected to job made by the welder.

Internal cost calculation can be used to arrive at the cost of welding a product, and also for comparisons between different procedures or for investment in new equipment, e.g. for the introduction of automation. It can also be used when evaluating various ways of reducing welding costs.

The model described here is best suited to ordinary fusion welding methods using filler materials. The calculations can be performed manually, or there are also more or less comprehensive computer programs that assist the work and so make it easier to compare a larger number of alternatives. Some programs include data bases for materials, filler materials, hourly rates, guide values for welding parameters and so on, which further assists the work.

It is important to realise that the accuracy of the calculations can never be better than the quality of the input data. Experience from earlier calculations, for which true costs have been obtained by post-production costing, is therefore important in ensuring that cost estimates are reliable.

23.2 Some welding cost concepts

Welding costing uses a number of concepts, as discussed below.

The *deposition rate* is the mass of weld metal melted into the joint per unit of time (while the arc is struck), and is usually expressed in kg/h.

In the case of welding with coated electrodes, the data sheets from the electrode manufacturers give the deposition rate at 90 % of the specified maximum current for the electrode diameter concerned. As the deposition rate depends on the current, it can be expressed in a simplified form for a particular value of current as

$$Actual\ deposition\ rate = \frac{Maximum\ deposition\ rate \times Welding\ current}{Welding\ current\ at\ maximum\ deposition\ rate}$$

The current depends on the welding position, whether the pass is for a root pass or a filler pass etc.

For MIG welding with solid wires diagrams and/or tables of deposition rates for different wire diameters and currents are available: alternatively, they can be calculated from the wire feed speed. In the case of steel, with a normal stickout, the wire feed speed is approximately:

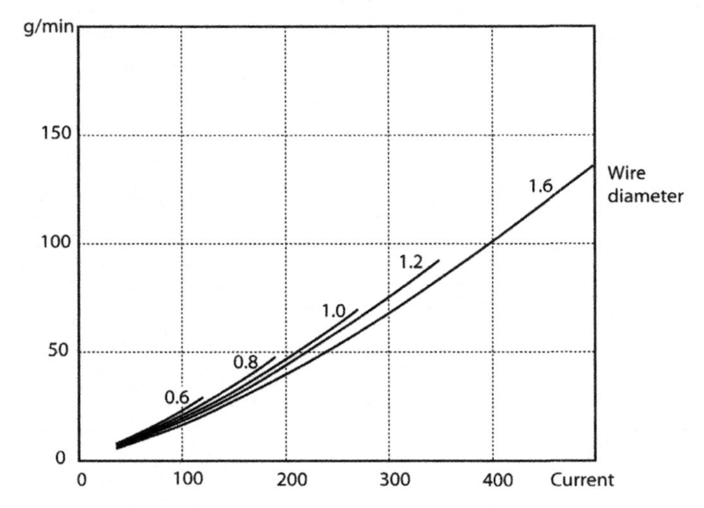

Figure 23.1 Deposition rates for ordinary wire sizes (steel).

The *deposition efficiency* indicates the proportion of the gross weight of filler material actually used that is converted to useful weld metal. Coated electrodes and flux-cored electrodes form slag, and so the deposition efficiency for such material is considerably less than 1. The stump length of coated electrodes is normally assumed to be 50 mm, except for stainless electrodes, for which it is taken as 35 mm. In the case of filler wire, there is some loss in the form of spatter. Typical values of deposition efficiency are given in Table 23.1.

TABLE 23.1 Typical values for comparative cost calculations.

	Operator factor	Deposition efficiency
Manual metal arc welding	0.30	0.60
MIG/MAG, solid wire	0.45	0.95
MAG, flux cored wire	0.40	0.85
Submerged arc welding	0.80	0.98

If the deposition rate is calculated from the wire feed speed, it is necessary to multiply the result by the deposition efficiency in order to compensate for slag or spatter losses.

The number of electrodes needed in order to deposit 1 kg of weld metal is of interest only for coated electrodes. If the weight of weld metal is known, the number of changes of electrode that will be required can be calculated and also the associated time for slag removal after each electrode.

The *weight of weld metal* is calculated from the details of the joint as shown on the welding drawings, which gives the volume of weld metal, multiplied by the density of the material which, for steel, is 7.8 kg/dm^3. The cross-sectional area of the weld, and its length, give the theoretical volume of weld metal. A fillet weld with a throat thickness of 4 mm and a length of 1 m has a weld metal volume of 16 cm^3. In practice, this value may need to be compensated for excess deposition, gaps etc., which increase the volume, and/or shrinkage, which decreases the volume. Tables are available that give the weld metal volumes for various types of joints. The previously mentioned computer programs automatically calculate the volume from the given joint data.

The *job time* is the time that it takes to carry out a particular welding job, and includes the setting-up time and the operation time. The *setting-up time* is the time required to set up and start the new work. In the case of mass production processes, this time will be incurred only in connection with the first workpiece to be produced. The *operation time* is the time taken for welding each workpiece, and includes:

- The *arc time*, i.e. the time for which the arc is actually struck.
- The *additional time* directly connected to the arc time, i.e. the time taken by replacement of the filler wire / electrode, slag chipping, cleaning gas nozzles etc., and directly related to the welding.
- the *handling time*, i.e. the time for handling workpieces, preparation prior to welding, such as tack welding etc. This time is very dependent of working practices and procedures.
- the *contingency allowance*, i.e. time that cannot be directly related to the welding, and which is often allowed for by a percentage addition.

The relative proportions of the total made up by the above times will vary, depending on the types of items produced, the way in which the work is carried out, the availability of ancillary equipment, the amount of mechanisation and so on.

The arc time is the time for which the arc is actually burning, and can be arrived at in various ways. Tables and diagrams from the literature and from suppliers give guide values that can be used in cases where a company does not have experience of its own to draw on. Own values, of course, have the advantage of applying exactly for the company's own particular situation and circumstances. In those cases where filler material is used, the *arc time* can also be calculated from the weight of weld metal divided by the deposition rate.

The *operator factor* expresses the relationship between the time for which the arc is struck and the total welding working time. It has sometimes been incorrectly used as a measure of productivity – i.e. assuming that a high operator factor indicates a high productivity. However, using a welding method having a higher welding speed will reduce the arc time. If other times are unchanged, this will mean that the operator factor is reduced, despite the fact that the work has actually been performed more quickly. If the company has experience-based values of operator factors for some particular type of welding, the total job time can be calculated by dividing by the operator factor. Typical operator factors are shown in Table 23.1.

23.3 Cost calculation

The *labour cost* is obtained by multiplying the *operation time* by the hourly cost. Companies know their own hourly costs, which are made up of hourly direct wages costs, employers' social insurance charges, holiday pay etc. The hourly cost often includes a share of common costs in the form of a percentage addition.

The cost of consumables needed for welding is referred to here simply by the umbrella name of *cost of filler materials.*

- Electrode (or wire in TIG welding) consumption can be calculated from the weight of weld metal divided by the deposition efficiency. Kilogram cost of electrodes is company-specific.
- Flux consumption is linked to weld metal weight by a specific flux consumption factor, stated by the flux manufacturer.
- Consumption of shielding gas can be calculated from the arc time (or corresponding time during which gas is flowing, e.g. as back shielding gas, multiplied by the gas flow rate in l/min. The gas cost (euro/l) depends on factors such as the method of supply/delivery (gas bottles, bulk tanks), and is specific for the company.

Machine cost can include costs for welding equipment, mechanisation equipment, special handling equipment and so on, and can be considerable if it includes automation and/or robot equipment. The hourly cost of the equipment can be calculated from the costs for depreciation, interest and maintenance, together with an estimate of the annual use time. The details of this calculation, and of the elements to be included, vary from company to company. The machine costs for a particular welding job can be calculated from the hourly machine cost and the operation time.

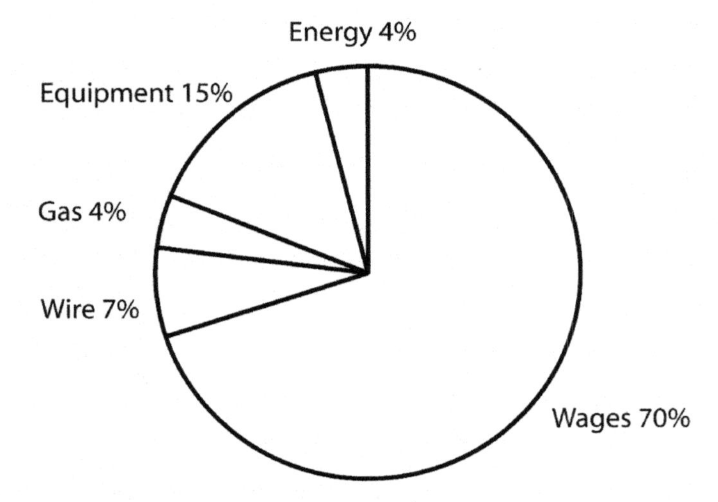

Figure 23.2 A typical cost distribution at manual MIGMAG welding (AGA).

The *energy cost* can be calculated from the arc time and power demand during welding, possibly with the addition of an allowance for no-load operation. It is not normally necessary to allow for the uncertainties that have to be considered for the other costs. Energy costs usually amount to about 1–2 % of the total costs.

Labour cost

Wages: $\dfrac{\text{Weld metal (kg) x hourly cost (euro/h)}}{\text{Deposition rate (kg/h) x Operator factor}}$

Cost of filler materials

Wire: $\dfrac{\text{Weld metal (kg) x Wire cost (euro/kg)}}{\text{Deposition efficiency}}$

Gas: $\dfrac{\text{Weld metal (kg) x 0.06 x Gas consumption (l/min) x Gas cost (euro/m}^3\text{)}}{\text{Deposition rate (kg/h)}}$

Flux: Rel. flux consumption (kg/kg) x Weld metal (kg) x Flux cost (euro/kg)

Backing: Weld length (m) x Backing cost (euro/m)

Machine cost

Equipment: $\dfrac{\text{Weld metal (kg) x Machine cost (euro/h)}}{\text{Deposition rate (kg/h) x Operator factor}}$

Energy cost

Electricity: Energy consumption (kWh) x energy cost (euro/kWh)

Figure 23.3 Formulae for calculating welding cost.

The cost of backing support is a function of weld length.

Comparison of all the above costs shows that is the labour cost that is by far the highest for manual welding. A typical example from welding plain carbon steel can be 80–90 % for labour cost, 5–15 % in cost of filler materials and 3–5 % for machine costs. The cost of filler materials becomes more significant when welding expensive materials such as stainless steel. Machine costs become significant only when automated and robot welding come into the picture. Measures to reduce costs are therefore concentrated on reducing the operation time.

Figure 23.3 shows a cost calculation based on theoretical calculations of the weld metal weight.

Throat thickness

The throat thickness in fillet welds decides the amount of filler material and also greatly influences the welding costs. If the penetration can be taken into account when the strength is calculated it is possible to save costs. The throat thickness can be calculated by the following relation.

$$a = d\sqrt{\frac{v_e}{v_l} \cdot \frac{k \cdot \pi}{4}}$$

where

a = Throat thickness
d = Diameter of the electrode
ve = Wire feed speed
vl = Welding speed
k = Deposition efficiency

Computer programs for welding cost calculations

Various types of computer programs for calculating welding costs are available. The simpler ones lack data bases of filler materials, additional times etc., but are cheap and user-friendly. At the other end of the scale is the Weldplan program (developed by the FORCE Institute, Denmark). It can be used to produce a complete specification of the work, which then forms the starting point for the cost calculations. Figure 23.4 illustrates a calculation using the program. The program can also produce an operations list, itemising operations to produce a total manufacturing cost.

Anderdahl Svetskunskap AB
BGT Components AB
268 33 SVALÖV

Cost Calculation

pWPS no: p013:00
Page 1 of 1
Date: 12/23/2002

Made by: Arne Anderdahl, EWE

Customer: Arjo   Project: Fästplatta 1-23949D   Place:

Preparation

	Time (h)	Cost
Gouging:	0.00	0.00
Backing:	0.00	0.00
Grinding:	0.00	0.00
Tacking:	0.00	0.00

Run	Electrode		Flux, Gas		Time consumption (h)			Cost Calculation			Total
	kg	Cost	kg	Cost	Arc	Process	Welding	Equipme	Welding	Per kg	Total
1	0.08		0.00	0.02	0.04	0.04	0.07	1.41	10.95	160.68	12.38
Total				0.02				1.41	10.95		12.38
Total costs (Prep. + welding)											12.38

Figure 23.4 An example of a welding cost calculation, using the Weldplan program.

Methods of reducing welding costs

One of the most important reasons for making cost calculations is to identify ways in which manufacturing costs can be reduced. Costs are influenced right from the design stage, with further input factors all the way through to production. Some examples are given below.

The biggest cost in manual welding is the *labour cost*. One way of reducing it is to introduce mechanisation and automation, described below.

The various parts of the total job time can be affected: a welding method with a higher deposition rate reduces the arc time, while changing the method completely might also make it possible to reduce the time needed for changing electrodes, slag chipping and spatter removal, thus reducing the total time. Equipment to hold or manipulate the workpiece to provide a good welding position assists welding. Planning of the work, too, is important, as perhaps only 30 % of the total time is productive arc time. It can sometimes be possible to avoid making unnecessary welds, and/or to use other production processes such as bending.

Some of the work time is the *arc time*. Even with a given welding method, it may still be possible to improve this by optimum choice of welding parameters and electrodes, and/or by avoiding depositing more weld metal than necessary. The design stage, for example, specifies the joint design and throat thickness of fillet welds. The joint design can be such as to minimise the amount of weld metal required, naturally subject to the necessary performance requirements. Too large a throat thickness always results in more weld metal than is needed: a throat thickness of 5 mm uses 56 % more weld metal than does a throat thickness of 4 mm. It is also important to plan joint preparations, bringing together and holding the parts and welding so that no more weld metal than necessary is deposited: this will also have the additional benefit of reducing welding deformation. If, in addition, the penetration of the fillet weld can be utilised to reduce the nominal throat thickness, there will be a further reduction in the quantity of weld metal.

The use of *filler materials* can also be influenced, although this cost needs to be related to the labour cost. If the labour cost can be reduced by more efficient welding, reduction of preparation and finishing, such as spatter removal, avoidance of two-sided welding, improved quality etc., additional cost for filler materials can be justified.

Bear in mind, too, the overall production process. The correct quality of materials, fillers, careful joint preparation and bringing together of parts all assist welding, reducing the overall time and having the least possible effects on other processes. A properly made weld generates fewer problems of inspection and possible corrections.

23.4 Mechanisation, automation, robot welding

Welding efficiency can be improved by the use of varying degrees of mechanisation. A certain level of mass production, or repeated production, is usually necessary, although various types of mechanisation equipment can be justified even with small batches. The introduction of mechanisation will alter the proportions between the various cost elements that make up the whole.

Setting up times are likely to be longer, particularly with more advanced automation. If manufacturing cost is to be viable, the cost of the setting-up times must be spread over a number of workpieces. The total savings emanate from lower operation time per item.

Each welder can also produce more through use of the equipment, although the machine cost will be higher, reflecting the use of the advanced equipment.

Quality control costs are likely to be reduced, as production is more carefully managed when the equipment has been correctly set up. However, training and running-in will be needed, which introduces extra costs. Maintenance costs will probably also increase, in connection with the use of more complicated equipment.

Properly installed and used mechanisation / automation / robot processes improve working conditions in both physical and work content terms. Better utilisation of capital bound in products in stock is another positive factor, resulting from the ability to tailor manufacture to suit actual needs.

Index

W

Y